America in the British Imagination

America in the British Imagination

Edited by

Catherine Armstrong, Roger Fagge and Tim Lockley

CAMBRIDGE SCHOLARS PUBLISHING

America in the British Imagination, edited by Catherine Armstrong, Roger Fagge and Tim Lockley

This book first published 2007 by

Cambridge Scholars Publishing

15 Angerton Gardens, Newcastle, NE5 2JA, UK

British Library Cataloguing in Publication Data
A catalogue record for this book is available from the British Library

ISBN 1-84718-169-4; ISBN 13: 9781847181695

TABLE OF CONTENTS

INTRODUCTION

CATHERINE ARMSTRONG, ROGER FAGGE AND TIM LOCKLEY

America has been a place of fascination for Britons for more than 500 years. The geographical location of the British Isles on the Atlantic fringe of Europe, and the disengagement from continental territorial conflict following defeat in the Hundred Years War in 1453, combined to turn British eyes westwards very quickly after 1492. Just five years after Columbus made landfall in the Caribbean, Henry VII sponsored John Cabot's voyage west to discover new lands and routes to Asia. News of his landfall somewhere on the Atlantic coast of Canada circulated widely amongst the literate and educated in England but much of the earliest information about North America was a mixture of fact and fantastical stories. Over the following century other explorers from Britain, Spain and France visited much of the coast and some of the interior of North America. As information about the continent gradually trickled into Britain and was translated, published, and re-published, myth gave way to slightly more credible facts, though still often embellished. As Britons learned of the freely available game in the dense woodlands, and the abundant off-shore fisheries, increasing numbers began to conceive of North America as a good place for overseas voyages and, ultimately, colonisation. It was not, however, until the 1580s that the first British attempt at colonisation in North America was made, at Roanoke Island in what is now North Carolina. The widespread circulation of drawings by artist John White and the accompanying descriptions of naturalist Thomas Hariot, made while the pair were in North Carolina, brought North America to life for those who had not gone to Roanoke. For the first time it was possible to read accurate descriptions of the flora and fauna of this part of North America and to see watercolours of the native inhabitants. However, the loss of the Roanoke colony after 1587, combined with the increased Spanish threat, deterred further attempts at colonisation until 1607 when a small band of adventurers settled on the James River in Virginia.

With the settlement of Jamestown, the amount of information about America flowing back to Britain increased dramatically. Not only were numerous promotional pamphlets published by the Virginia Company, which often needed to be taken with more than a pinch of salt, for the first time negative images of America began to appear in England complaining of the harsh climate, the unfriendly Indians, and the brutal discipline regime of the colony's leaders. The four hundred years following the settlement of Jamestown have continued to see the British assailed with two quite distinct images of America. On the one hand America is a land of personal freedom and individualism, of boundless economic opportunity, and broad social equality. On the other hand America has been perceived as having a self-confidence bordering on arrogance where higher culture is usually subsumed by materialism, where the poor can be easily left to fend for themselves, and where foreign policy can be dogmatic and aggressive. One theme of this book is that Britons do not know whether to envy America or revile it, but one thing is certain—they do still think about it. No other nation on earth is quite so prominent in the minds of the British. In part this is because America offers to the British a number of "alternate universes" where the effect of social experiments can be observed. The British migrants to the American colonies brought with them the English language, Protestantism, democracy, the rule of law, the separation of powers, and economic liberalism which led to an underlying similarity with the mother country. However, no American colony was identical to Britain. Nowhere in Britain were religious leaders so socially and politically prominent as they were in seventeenth-century New England, Pennsylvania or Maryland; slavery, which came to define the southern colonies during the eighteenth century, had no counterpart in Britain; and, since land was far more readily available to migrants in America than back home, franchise requirements were more easily met meaning that in most colonies a far higher proportion of adult white males were permitted to vote. These differences meant that Britons back home could gauge the effect of each of these changes on an essentially British society. If they wanted to see what a British society with slaves would look like, for instance, they needed to look no further than South Carolina. Following the American Revolution, Britons were also able to see how a "British" society would look like without a monarchy or an aristocracy.

The popular democracy and perceived social mobility of post-Revolutionary America was deeply attractive to many Britons, and the rekindling of cultural and personal ties between America and Britain in the nineteenth century ensured that both continued to exert an influence over the other. The causes of abolition and social reform, for example, were fought on both sides of the Atlantic. Britons continued to visit America, and were fascinated and repelled by the young nation. Did it offer solutions to the social problems experienced in the world's first industrialised nation? Could it offer alternative modes of government and social organisation to put an end to poverty, criminality, and conflict? The answers, such as they were, varied from the impractical to the implausible.

Towards the end of the nineteenth century America began to take its place as an imperial power with new colonies in the Caribbean and the Pacific, but just as America's world influence was growing Britain faced Armageddon. The First World War irrevocably changed the respective positions of Britain and America. Instead of being a great imperial power, Britain was about to embark on the long retreat from an empire it could now ill-afford. America, flushed with the success of its decisive entry into the war in 1917 emerged as a great power in its own right. British views of America also changed as self-confidence in the British way of life gave way to self-doubt. Perhaps America really did offer a better alternative.

World War II confirmed America as the world's leading industrial and military power, and the post-war period has seen no diminution of British interest in America. American popular culture, especially film, music and television, became almost universally known in Britain, even if reviled in some quarters, but it is American foreign policy during the Cold War and afterwards that has attracted the most British interest and criticism. As the sole western superpower America launched itself into overseas ventures in Korea, Vietnam and most recently the Middle East. Undoubtedly, British perceptions of America have been coloured by these conflicts.

This volume emerged from a one-day conference co-sponsored by the United States Embassy in London and the University of Warwick, and was conceived as an opportunity for a number of scholars from a variety of different disciplines (History, English, Theatre Studies, Music and History of Art) to explore the ways in which Britons have imagined America. The time frame runs from the seventeenth to the twenty-first century, and should enable the reader the see how British perceptions and understandings of America have evolved over those 400 years. Many essays build on recent scholarship by David Armitage and others around the concept of the "Atlantic World." Rather than seeing the relationship between former colony and mother-country as based on simple bi-directional power-relations, this collection aims to foster a sense that these links were complex and multi-faceted. It also acknowledges that events in Europe, Africa and the rest of the Americas influenced the ways that the British imagined the United States. Connections between these two countries must be explored by looking at literature, theatre and music as well as the political and economic connections, here placed into a more flexible context, not presented as deterministic drivers of, say independence, or the Cold War.

Catherine Armstrong's essay starts the volume by highlighting the breadth and sheer pervasiveness of printed representations about America before 1630. The print media was not simply confined to travel narratives and descriptions, which were often lengthy and required higher-level literacy skills, but also much shorter ballads, plays, poems and sermons. These items sometimes required only basic literacy, and could be transmitted orally. Armstrong shows that the Virginia Company deliberately used these informal methods of communication to ensure that news of the settlement at Jamestown reached the widest possible audience, and especially those among the lower orders who were potential colonists. Since the Virginia Company were responsible for many of these printed representations they were normally very positive about the new colony but gradually they lost their monopoly on the flow of information from Virginia. Negative reports were widely circulated, especially following the starving time of 1609-10 and the 1622 Indian massacre, and these were extremely influential in shaping the future image of America in the British mind.

Sarah Irving's study of the Royal Society demonstrates how information about America was deeply intertwined with ideas of empire, dominion and control. Natural scientists who were members of the Royal Society tried to place America within existing schemes of classification whilst also making the conceptual leap that "conquering" the New World in these terms helped to "conquer" it in reality as well. Since man's natural sovereignty had been divinely ordained before the fall from Eden, understanding, labelling and classifying new flora and fauna, helped Britain simultaneously to claim sovereignty over new territory. In this sense, only by knowing about the new world could the British truly conquer it. Ian Chambers takes up this theme of the power of words and naming in the eighteenth century in the first essay in this volume to explore how the British imagined the native inhabitants of the New World. He demonstrates, using Scotsman Alexander Cumming as an example, how British and Native American understandings of the same events were radically different: British travellers thought they were securing the loyalty and submissions of southern Native Americans but tribes such as the Cherokee saw it far differently. However, far from returning back to Britain with a corrected understanding of tribal diplomacy, cultural misunderstandings were often perpetuated and reinforced because of the ignorance and closed-mindedness of travellers. Popular British understanding of Native Americans can perhaps best be described as muddled and inaccurate.

John Barrington's study of the mentality of British settlers in South Carolina is the subject of chapter four. Barrington suggests that these "British Americans" viewed themselves as an integral part of the British Empire, not just in political terms but also because of a shared determination to prevent French/Catholic hegemony in the New World. "British Americans" in South Carolina saw America as vital for British imperial strength partly because of the raw materials that eased reliance on important European goods, but more particularly because in the eighteenth century the war between France and Britain was actually being fought in America. In a pre-cursor of the cold-war ideology of the late twentieth century, a strong America meant a strong Britain.

Moving into the nineteenth century, Robert Grant uses the visual representations of a newly independent America and compares them with those of Canada, New Zealand and Australia to show how the British conceived of their current and former colonies as new worlds, or "tabula rasas," that were simply waiting to be moulded into a proper form by colonists. All new lands had both positive and negative features: the ability to create order from chaos and fashion a new civilization was balanced by the hard work involved and sometimes harsh natural environments. But America, and all the new British colonies, formed part of a larger, Anglo-centric, mentality of superiority and dominance over others.

Max Stites uses the work of novelist Charles Dickens to explore British views of post-Revolutionary independent America. Dickens, like many of his contemporaries, perceived America as a place of opportunity and freedom, but ultimately he was forced to accept that the reality was somewhat different. His personal experience of America complicated rather than confirmed his preconceived ideas and the harsh and competitive nature of existence in the New World became a theme of several of his fictional works. Stites argues that Dickens ultimately conceived of America as "amputated" from Britain, with the underlying suggestion that perhaps America would have been better served by remaining part of the British Empire instead of striking out alone.

David Brown and David Waller use the reports of British Consuls based in southern ports to contradict long-held views about British sympathy for the confederacy. Whilst the political elite in Great Britain was increasingly suspicious of American ambition, especially in the west, and some recommended support for southern independence simply to weaken the USA, ultimately the issue of slavery got in the way. The South was seen to be a deviation from the Anglo-American norm, and indeed many believed that the attachment of southerners to the institution of slavery was a failure of popular democracy. The shared political ethos with northern states, and a recognition that the North, more than the South, had grown from British racial stock, ultimately meant that Britain increasingly turned against the confederacy as the war progressed.

As the United States grew economically in the latter half of the nineteenth century it increasingly clashed with the world's dominant economic power—Great Britain. Stefan Schwarzkopf's essay explores the ways in which the commercialism, entrepreneurship and showmanship of American advertisers won them increased market share ahead of their more reserved British counterparts. In Britain this was often perceived as an example of the shallow and degraded nature of American society, with British advertisers preferring to stress the quality of their products and to establish trusting relationships with consumers. While some in Britain believed that since American techniques were clearly successful they should be emulated, many others thought they had little, if anything, to learn from America. The different approaches adopted by American and British advertisers therefore acted to accentuate and highlight trans-Atlantic differences between these two economic superpowers.

By the late nineteenth century some British authors were acutely aware that America was catching-up to Britain economically, and would inevitably overtake her. Brook Miller uses the travel writings of three British visitors to the United States to show how cultural critiques of America were, in large part, more about the defence of British culture than about a denigration of America per se. Authors shored up British self-esteem in two distinct ways: firstly they emphasised the close ties between Britain and America, suggesting that American dynamism was mainly because of the nation's British origins; but secondly they argued that regardless of the pre-eminent economic position that America would ultimately take, the absence of a collective history or identity, and the homogeneous nature of America's "classless" society which lacked any real "refinement" meant that it could never supplant Britain culturally. This smug sense of cultural superiority was to become a central theme in British feelings towards America in the twentieth century.

Cultural superiority was not totally absent from the radical British critique of America either and, as Roger Fagge shows focussing on the 1930s, the left were unsure how to deal with the industrial-capitalist giant across the Atlantic. On one hand they dismissed it on ideological terms, but on the other their personal connections with the country, and the progressive politics of the New deal, encouraged a more subtle response. This was particularly true of the cultural radicals H.G. Wells, J.B. Priestley and W.H. Auden who all embraced elements of the American experience. Even for these radicals however, America remained something of an enigma, which they tended to re-imagine to fit their worldview.

Allen McLaurin turns to the turbulent decade of the 1940s, where the pre-war admiration for America in some quarters gave way first to the close alliance of the Second World War and then to the post-war world that seemed to confirm British global decline and American pre-eminence. Some writers of the later 1940s indulged in a resurgent anti-Americanism and a re-assertion of British cultural superiority where Britain played the part of Greece to America's Rome, while others continued to praise American inventiveness and energy. McLaurin ultimately suggests that writers of the 1940s had a highly ambiguous view of America.

Novelist Kingsley Amis, as Tammy Grimshaw explains, was one of the first writers to cast a satirical eye on American self-perception. Amis's works of 1960s often use comedy to puncture the notion that American foreign policy was distinctive from that of European powers. In fact nineteenth-century territorial growth, both on the mainland and overseas, was deeply imperialistic and, unlike Britain, America was still engaged in imperialism in Asia in the twentieth century. Amis was sharply critical of the Vietnam War for instance, and much of his work has an anti-American tinge. Yet, as with so many Britons who feature in this collection, Amis denied that he was anti-American. His satires of American morality and racial attitudes therefore reflect a typically British confused and enigmatic attitude towards the United States.

During the early 1970s American playwright Sam Shepard took up residence in London and wrote and produced a series of works on American themes. Tim Rogers' essay examines the techniques used to promote Shepard's work to British audiences, and in particular the way in which Shepard's own disillusionment with America led to his depiction of his native land as a foreign, alien environment verging on the grotesque full of strange people. Building on this sense of disappointment with America is Joseph Tate's essay on the band Radiohead. Tate brings us into the twenty-first century by using the song lyrics, as well as the public statements of the band members, to show how Radiohead have consistently attacked American foreign policy. These musicians imagine an America that is a threat to world peace, and no longer the force for morality and right that it had once been.

The widely differing essays in this book cannot hope to be comprehensive. At most they can only provide a number of intriguing snapshots of the ways in which Britons have imagined America from the seventeenth to the twenty-first century. A common thread is the confused nature of British imaginings. America is both "the land of the free" and "the great Satan" and British people can love America both as a concept and as a place to visit, while reviling the American government and especially its foreign policy. We hope that these essays will stimulate future scholars to continue to reflect on the ways in which Britons have thought about their most significant external influence—the United States of America.

CHAPTER ONE

CONTESTING THE MEANING OF AMERICA: PRINTED REPRESENTATIONS BEFORE 1630

CATHERINE ARMSTRONG

In the first hesitant years of exploration and settlement how did English men and women acquire and interpret knowledge of the American colonies? This chapter will focus mainly on Virginia, which, before 1630, featured predominantly in both positive and negative reports, although during the 1620s some news was beginning to emerge from the Plymouth colony to the north. In the early years, English readers were intrigued with Virginia and the Caribbean colonies and as a result popular demand dictated that most of the news coming from America concerned these areas. However it is not my intention to try to interpret the views of the first white settlers in the region, those early colonists who remained in America, but to look at the transatlantic view: the understanding of America from the shores of Britain. Many of the authors were colonists disillusioned with the New World who wanted to share their experiences with others on their return to England. Some were explorers who had only intended to spend a few months or years travelling, surveying and claiming new territory. Others who wrote about America never left England's shores, indicative of the pervasiveness of knowledge about America, that it was firmly entrenched in the hearts and minds of many people, inspiring them to put pen to paper to spread news about the colonies. Many of these men intended simply to tell a story, to place themselves in an heroic or tragic narrative, while others wanted to use tales from America as a sounding board for the creation of hopes of a new English empire. The two great editors of travel narratives, Richard Hakluyt and Samuel Purchas, both believed that by encouraging information about English exploits overseas to seep into the English consciousness it would ferment a desire among their countrymen to see the colonies succeed.[1] They believed, Purchas especially, that the success of Virginia depended not only on meeting the financial imperative of creating a self-sustaining colony, but also on Protestant evangelising. When Purchas described Englishmen travelling overseas as "pilgrims," he was not harking back to the days of Medieval spiritual journeys, nor was he referring to the exiled, separatist Englishmen who ended up at Plymouth Rock in the 1620s, but rather extolling his readers to become Protestant messengers, by investing in Virginia as well as migrating there. The complexities of pursuing material and spiritual gain simultaneously caused problems for the ministers who wrote about America. The literary ways in which they extricated themselves from this philosophical dilemma gave rise to the earliest expressions of manifest destiny, the belief that God intended the land of North America to belong to the English. This had a profound effect on the relationship between colonist and native. It negated the role played by native inhabitants in supporting the colonists in the first years of settlement and deprived the natives of control over their own destiny.

Introduction: Printing and Adventuring

This chapter not only encompasses the printed representations of America but also examines the media through which these representations were defused. Some books about America were distributed during Elizabeth's reign, many in Spanish, being derived from the explorations of that nation in Central and South America. These works aroused the interests of a few English traders who were involved in the supply of slaves and purchase of goods in that region. But most readers only really showed an interest when the English became involved in exploration and settlement in that hemisphere. The stories of the earliest English explorations in North America were crafted as logs and diaries by the sailors on the voyages of great captains such as Drake, Frobisher and Gilbert.[2] However, when these narratives were gathered together and published by Richard Hakluyt in printed form, knowledge of America began to spread much faster, firstly at court, then among other members of the elite.[3] At that time the poorer members of society had little access to this information because Hakluyt's volumes were very expensive, even when sold unbound, so they were affordable only to the wealthier reader. By looking at the practicalities and technicalities of production and distribution of news in London and the provinces, it is possible to say a little more about the role of the authors, playwrights, printers and booksellers in creating an awareness of America.

It is also important to consider why authors and producers of books and pamphlets about early America chose the medium of print. Perhaps they were answering John Donne's exhortation in a sermon of 1622 that "every man that prints, adventures." Donne was a keen supporter of the Virginia Company and was its secretary for a few years. He gave sermons to the gathered Company on several occasions and was certainly aware of the importance of employing the medium of print to promote the colony. However, the picture is more complicated than that. Much of the information from the colonies reached England in manuscript form, with the author unaware that his work was to be turned into a printed tract. The Virginia Company, led by Sir Thomas Smith encouraged this enterprise, just as it attempted to suppress those letters and diaries containing bad news about America. Even John Smith, the most successful author to write about the New World in English during this period, had his first tract published without his knowledge, the anonymous editor apologising in print in the second edition for not acknowledging Smith as author in the first.

The distribution of printed material about America centred on London, as did most of the printing trade. The presses of Oxford and Cambridge were active by this time, however all but one work on America was published in London. John Smith's *Map of Virginia* was the exception, being published in Oxford in 1612; some have argued this was because William Symonds, a doctor of Divinity at Magdalen College, contributed a prefatory section, while others argue that the book was suppressed by the Stationers' Company and was forced out of London. Whatever the truth, the fact remains that most works at this time were being produced in London. The printing and bookselling trades were centred on St. Paul's Churchyard, with the more successful tradesmen having shops in the area, and other smaller traders selling from stalls or even wandering the streets. These pedlars often took the smaller books outside London to market towns in the provinces where interested parties read broadsides and small pamphlets to gathered crowds, thus bringing information about America to more than those previously defined as truly literate.[4] Members of the wealthier classes in the countryside bought books from the London shops through dealers, or attended specialised book fairs held around the country.[5] Sadly, catalogues from these early seventeenth century fairs and book agents do not survive, although examples from the late seventeenth and early eighteenth centuries reveal a strong interest in books on travel, geography and history.

However, not all authors and tracts were seeking to encourage migration to, and investment in, Virginia. From the earliest days of settlement in Elizabeth's reign, when the inhabitants of the small English colony at Roanoke vanished, presumed massacred by the natives, rumours spread of the terrible conditions faced in the New World and a counter-narrative of fear and disappointment was born. In the early seventeenth century, this negative response was often the result of cynicism caused by the over-enthusiastic lauding of the enterprise. Playwrights such as Ben Jonson were cutting in their satirical portrayals of the supposed wealth and security of the New World. The interaction of these two narrative strands allowed print culture to affect the ways in which the English thought about North America.

Part 1. Positive Reports

To those who spent a very short time in America, it inevitably seemed like a second paradise. This was partly because of the traditional sailing times of ships from England; they would arrive off the coast of Virginia and New England in late summer and early Autumn, with their sailors and passengers seeing the abundance of flora and fauna without experiencing the extremes of heat and cold that were present in the region in high summer and the depths of winter. John Brereton recorded the voyage captained by Bartholomew Gosnold in 1602, which was sponsored by Walter Raleigh who vainly hoped to discover his lost colonists from Roanoke. Its other stated aim was also unsuccessful, that of revealing a naval passage to the South Seas through which the English might access the lucrative spices and silk trades of South Asia. However, these failures did not prevent the expedition from exploring the coast of North America and engaging with the natives who Brereton described as "of a perfect constitution of body, active, strong, healthfull and very wittie," primarily because of the "holsomenesse and temperature of this climat."[6] Brereton thought the prospect for an English colony if settled there proved good, "our habitations will be made a staple of all vendible commodities of the world," especially as "no nation of Christendom is so fit for this action as England." The natives, though healthy, provided no threat, for they were 'simple, naked and unarmed."[7] Although written later, when the colony of Virginia had been established, the report from Christopher Levett's voyage of exploration sounded a similar note to that of Brereton's twenty years earlier. Levett also explored the area known for many years as Northern Virginia, but by the time of his journey named New England, and reported that "a good plantation may be settled, for there is a good harbour, good ground and much already cleared for plantinge of corn and other fruits."[8] Even the early report of John Smith from Virginia portrayed a peaceful existence, despite the tensions that were already building up among different factions of settlers. Smith concluded his 1608 tract on Virginia with the words "we now remaining in good health all men well contended and in love one with another…to set our Nation to enioy a country not onely exceeding pleasant for habitation but also very profitable for commerce in generall."[9] However, this enthusiastic trumpeting of the virtues of the New World aroused the suspicion of writers such as Jonson who thought it too good to be true, and it also encouraged migrants to travel to America unprepared for hard work and harsh weather, because they had been led to believe that no work was needed to till the soil and produce food.

Between 1609 and 1620 the most consistent publicist of the Virginian colony was the Virginia Company itself.[10] This should be of no surprise, as its members, already men of wealth and influence, hoped for further financial gain by using the medium of print to persuade their countrymen to invest or migrate, thus ensuring a self-sufficient and independent colony. The Company knew their audience well; sermons were preached to gathered members of the Court and Company in London to encourage them to keep up hope and maintain their financial interest. These sermons were then published. Migrants were also needed from the middle and lower orders of the society who could not be reached through grand sermons in London. Their attention was attracted through the medium of the broadside, a one-page news sheet, easily and cheaply distributed around the country by pedlars and booksellers. The first of these broadsides in 1609 advertised for "Artificers, Smiths, Carpenters, Coopers, Shipwrights, Turners, Planters, Vinears, Fowlers, Fishermen, Mettel-men of all sorts, Brickmakers, Brcklayers, Plowmen, Weavers, Shoomakers, Sawyers, Spinsters, and all other labouring men and women that are willing to goe to the said plantation."[11] Later, because many migrants had arrived in Virginia totally unprepared to support themselves until the next harvest, the Company issued another broadside, again encouraging migration, but this time listing the belongings that such a person should bring with them, including "Apparell … Victuall … Armes … Tooles … Houshold Implements."[12] Other broadsides were issued in connection with the Virginia lottery, probably only the third public lottery in English history. The company was granted this privilege by the king when it became clear that the colony was struggling to finance itself. Before the lottery was drawn, several broadsides were issued promising investors that their money would go towards the stocking of supply ships, which would relieve the "adventurers to Virginia [working] for the establishing of the Gospell there and the honour of our King and country."[13] The Virginia Company also showed awareness of another form of print culture: the ballad. Ballads were often thought to be an ephemeral form, but have recently been shown to be a valid expression of the political culture. The Virginia Company used a ballad to raise awareness of their lottery, targeting not only the middling orders of society who might have been interested in investing in the colony, but also the probable migrant labourers from the poorer classes. The ballad, entitled *Londons Lotterie*, was released in 1612 in two parts, each with three rough woodcut illustrations. It emphasises the honourable nature of the enterprise, that the colony would be of benefit to the Christian world, and to the converted "savages," and would bring honour to the King and Queen and Prince Henry. But the ballad also unashamedly appeals to those hoping to win the lottery, even to women:

> "You maydes that have but portions small
> to gain your marriage friend,
> Cast in your Lottes with willing hand,
> God may good fortune send.

You widdowes, and you wedded wives,
One little substance try:
You may advance both you and yours,
With wealth that comes thereby."[14]

The ballad also speaks to "farmers and countrymen," extolling them to remember the times of Queen Elizabeth under whom the colony was originally founded and given the name of "Virginia." It concludes by arguing that Englishmen would do well to remember that England, too, had been "a wildernesse and savage place" once and so it was possible to predict a transformation of Virginia into a flourishing and lively settlement.[15]

A more sober form of printed medium, derived from the sermons preached by ministers in London, was also employed by the Company to target those who had invested in the colony or who might want to invest in the future. Some work has been done on the significance of these sermons and their encouragement to listeners to fulfil the values of Humanism as well as of the Protestant Church.[16] But sermons should not be seen as an individual genre, rather they complemented other Company publications and other printed artefacts such as laudatory poems, economic tracts of encouragement to merchants and the output of so-called "armchair travellers." Sermons had the double impact of being powerful oratory, influential to those who heard the delivery, and to the reader later when they were preserved in print, usually by printer William Welby, who worked for the Company for several years. The sermons form two distinct groups: from the earlier period, around 1609, when the colony was founded and the company hoped to encourage more financial support, and then in the 1620s, when the settlement's future looked a little more secure, but was still threatened by instability. William Crashaw, speaking in 1609, in a sermon published, according to the title page, "without his consent," argued that the sole purpose for settling Virginia was to convert the natives to Christianity. He was speaking to the gathered throng of the Company including its governor, Thomas West, Lord De La Ware, who was about to embark on his first voyage to Virginia. His was to be the ill-fated voyage shipwrecked in the West Indies that influenced Shakespeare's play *The Tempest*. In a reversal of the reality in America, Crashaw naively suggested that the English settlers could assist the natives and teach them how to cultivate the ground for profit.[17] Even as early as 1609, Crashaw was able to identify the true enemies of Virginia: "the divell, Papists and Players," revealing that negative rumours about America were already being spread, with the theatre as a hub. William Symonds, speaking to the same audience, also in 1609, also chose to focus on the fear of Papist intrigue in the New World. He argued that the English must send as many men as possible to the region because otherwise the Jesuit missionaries would impose their doctrines on the Natives. As Symonds put it "it is a shame that the Jesuits and

Friars that accompany every ship should be so diligent to destroy souls and wee not feed the tender lambs, nor bind up that which is broken."[18]

The tone of the sermons in the 1620s was very different. They were preached in the context of increasing wealth and profit for colonists and Company, but this had come at a price of worsening relations with the Natives. Patrick Copland, whose involvement with Virginia led him to raise money for a school to educate the natives, preached a sermon before the Company in April 1622, just a matter of weeks before the news of the Virginia massacre reached London. One third of the English settlers were killed in the massacre, which was a co-ordinated attack by natives, many of whom had breakfasted with their victims a few minutes earlier. Copland, unaware of this tragedy, but obviously keen to contradict rumours of unrest, argued that there was "no danger after their landing, either through warres or famine or want of convenient lodging."[19] By the time John Donne preached his famous sermon, on the eighth verse of the first chapter of the Acts of the Apostles to the Company in November, the sad truth had reached London, and the recriminations and acts of revenge had already started. However, Donne preached that it was still possible to live alongside the Indians, converting and not killing them. He encouraged patience saying "you cannot beget a sonne and tell the mother I will have this sonne born within five months."[20] He concluded with what could be read as a subtle attack on the way the Company operated, arguing that they must not create an elite and that merchants and gentleman of the colony must not be at odds with one another.[21]

It would be a mistake to think that promotional material emerged only from the Company itself. The supporters of the Company's enterprise included some other illustrious poets apart from Donne. Around 1619 Michael Drayton wrote a famous ode "To the Virginian Voyage," in which he lauded the settlers who had gone to the New World as English heroes, encouraging others to follow their example:

> "Britons, you stay too long
> Quickly aboard bestow you…"[22]

Interestingly Drayton, like Donne, understood the importance of print culture to the Virginian enterprise, acknowledging in his last verse the role played by men such as Richard Hakluyt who championed the funding and migration central to the enterprise:

> "They Voyages attend
> Industrious Hakluyt
> Whose reading shall inflame
> Men to seek fame

And much commend
To after-times thy wit."[23]

No less laudatory, but written in the aftermath of the Virginia massacre, Christopher Brooke's tragic poem lists the fallen heroes of the enterprise, especially the captains who had led the colony. Brooke takes a different approach in verse to that of Donne in prose, recommending, not that the English might still live alongside the natives, but that the natives must be exterminated as they were "rooted in evill and opposed in good, errors of nature, or inhumane birth."[24] Brooke's tragic poem had an optimistic note however, praising the virtues of the new governor Francis Wyatt and exhorting him to wield the "sword and the pen" and govern the colony well to ensure its strong future.

Brooke's predictions seemed to come true in the following few years. The relations between the Europeans and the Natives in the region were permanently soured by the violent reprisals and a withdrawal of the English from communal to isolated living. But the colony was not abandoned and it continued to flourish, in the most part because of the burgeoning tobacco trade. This led to a new genre of literature praising the virtues of Virginia: the economic tract, written to encourage merchant activity in the region. The authors of these tracts, such as John Hagthorpe, who wrote *England's Exchequer* in 1625, had never been to the New World. Hagthorpe saw his book as a noble endeavour to encourage the English to emulate their European neighbours, especially Spain, and bring wealth to the nation through overseas trade. He also employed classical metaphors to emphasise the point, arguing that the flourishing civilisations of the Carthaginians, Romans and Greeks travelled to seek out new lands and goods to trade. However, the English mercantile endeavour would benefit not only those who sought to invest in the scheme, but also would be a relief to poverty by employing the poorest men in worthwhile labour. Hagthorpe recommended that this trade "seems rather a worke for the Prince and Commonwealth than for private men." In this, he was ahead of his time, as England's first monarch to take a direct approach to the colonial enterprise was Charles II. With no fear from Spanish attack, Hagthorpe saw the potential for trade with New England and Newfoundland, especially praising the latter whose climate he optimistically claimed to be ideal for the English temperament. He wrongly believed that Newfoundland would be protected from the threat of pirates because of the presence of the English fishing vessels.[25] This was not the case; Newfoundland was constantly plagued with pirates, some of whom kidnapped settlers and fisherman who were standing on the shore. Factual errors of this sort were frequent in tracts written by those who had never been to America. They wished to sound positive about the opportunities and benefits, but sometimes, in their enthusiasm, made false claims about the climate, the flora and fauna or the native populations. These enthusiasts, who might be termed "armchair travellers" were men who felt it was their duty to encourage others to invest in the enterprise or to migrate to America while they themselves could not go.

Richard Eburne, the vicar of Henstridge in Somerset, is a good example of an armchair traveller. He was fascinated by the colonisation of Virginia, but the benefits of Newfoundland really aroused his interest following the publication of Richard Whitbourne's 1620 tract *A Discovery and Discourse of Newfoundland*, which the Privy Council instructed to be promoted from every parish pulpit. The ministers of the church were often less interested in the financial benefits that America had to offer, but instead focussed on the curing of society's ills. As Eburne argued, America could provide "room for our overswarming multitudes of people."[26] The fear of over-population, and especially of large numbers of poor people was ever-present in Elizabethan and Stuart England and the colonies were seen as a double benefit because they might relieve some of the population pressures while putting the poor to work. However, Eburne accused the English of "wanting industry" and having an "immoderate love of their country," intending his tract to be part of the effort to encourage people to participate in the colonial enterprise. Like Hagthorpe, he argued that the King should take a more direct role and declare himself to be King of the entire region to be settled.[27]

Supporters of the colonies like Eburne usually favoured one region of North America over the others and their arguments are thus biased towards one colony to the detriment of the rest. Eburne felt that Newfoundland would be the best site for a socially improving colony such as the one he envisaged, although he did acknowledge that a purely money-making venture would probably be better located somewhere with mineral wealth such as Guiana. Another minister, the Reverend John White of Dorchester was involved in financing and publicising the voyages to Massachusetts in 1629 and 1630, and wrote a tract defending the enterprise, although he modestly claimed in the epistle to the reader that the work "has been wrested out of the authours hand, hardly overlooked, much lesse filed and smoothed for the presse was never intended to be presented to publicke view."[28] White was a godly Puritan, and concurred with John Winthrop's aim of creating a "city upon a hill" that would act as an example and a beacon to England to lead her through those godless times. White emphasised the redeeming nature of the overseas settlements; putting the unemployed to work and spreading the reformed word of God, and he reiterated the idea of manifest destiny, that God ordained New England as fit and right for English settlement because "the Lord affords void ground enough to receive more people than this state can spare," and that this had been achieved by "a defoliation" which killed all the natives. White was also aware of more earthly matters, and defended New England against accusations that its winters were too cold ("the cold of winter is tolerable…and is remedied by an abundance of fuell") and that it is infested with mosquitoes, but not as badly as Essex or Lincolnshire.[29]

After the flurry of interest caused by the massacre in 1622 had subsided, the works of Eburne and White reflect the movement away from Virginia in the literature on America. However, one armchair traveller, who probably did more than anyone in the 1620s to awaken interest in English exploits abroad, always saw Virginia as the jewel in the English colonial crown. Samuel Purchas gathered together many letters and journals recounting England's expeditions around the world and published them in his twenty volume collection *Purchas, His Pilgrims*. He also included much of his own opinion as in the tract entitled "Virginias Verger" in which Purchas passionately pleads with his readers to keep faith with the colony, which he argues is even more justifiably English following the "perfidiousnesse" of the Indians during the massacre.[30] He admits that Virginia has been a victim of malicious gossip, but he reiterates the reasons why Virginia should be considered a suitable venue for "evacuation" of the excess population of England. The potential for trade is also highlighted, with, according to Purchas, iron ore, timber, sugar and tobacco being the most fruitful. He also keeps alive the hope of finding a North-West passage to the South Seas, thus opening up a route to the East Indies for the English. However, Purchas also foresaw a great age of seafaring in which Virginia would be a friendly, safe haven for English shipping in peacetime and a useful base from which attacks against the enemy could be launched during wartime.

Part 2: Negative Reports

Purchas, writing in the mid-1620s, was fully aware that it was not only the damage caused by the massacre, but also other negative reports that had sullied Virginia's reputation. The first disaster to strike the colony was the difficult winter of 1609-10, known as the "starving time." Having failed to become self-sufficient and still unable to provide all their own food, the English settlers demanded and stole food from the natives causing them to withdraw the friendship that was helping to sustain the colony. Since many of the letters containing information about this difficult winter were sent to Company officials, most of them were not published. However, their circulation in manuscript and rumours spread by word of mouth proved enough to turn some English people against the enterprise. On 7 July 1610 a letter was written by Lord De La Ware, governor of Virginia, to the Earl of Salisbury in which he described Jamestown as a "verie noisome and unholsome place, occationed much bie the mortallitie and idlenesse of owure people." He thought the colony "never so weake and farr out of order as nowe," but even then sounded an optimistic note that given time the situation could improve, and if God would grant his good health he would undertake that change himself.[31] Samuel Purchas elaborated on the problems faced by the Virginians during this period, but he wrote with hindsight, emphasising that this phase had now passed. He wrote that "life was now preserved by roots, herbs, acorns, walnuts, berries and now and then a little fish. A report was that one slew his wife and had eaten part of her: this is by others denied." Purchas emphasised that the problem was caused by the settlers rather than anything lacking in the country itself: "The occasion of these miseries was onely our owne for want of providence, industrie and government and not the barrennesse and defect of the country."[32]

The colony managed to survive that difficult winter but its future was still uncertain until John Rolfe's cultivation of tobacco and its establishment as the staple crop ensured the survival, if not the stability of the colony. Tobacco was a controversial crop in the seventeenth century, as it is today, and many of Virginia's early critics were opposed to the choice of tobacco as its staple crop. Richard Eburne, although a loyal supporter of the colonial enterprise, railed against the "tobacco shops that now stink the land all over."[33] James I's own objections to tobacco smoking are well documented, but in 1616, John Deacon went even further, arguing that smoking had caused the treachery of the gunpowder plotters.[34] Its detractors regarded the practice of smoking, derived from the Native Americans, as "paganish and heathenish." In an era when many medical tracts claimed that tobacco was a panacea, Deacon sounds surprisingly like a twenty-first century commentator, arguing that "such excellent personages have not onely been sodainly surprised by unnaturall death but their dead bodies being opened had all their entrails as blacke as a coale."[35] Tobacco was thought to encourage men to drink wine and beer to excess and was an extra drain on their resources, with users "falling first from tobacco to tipling, from tipling to whoring."[36] Deacon drew a portrait of a tobacco smoker who neglected his wife and children and spent their maintenance on his filthy habit. In his conclusion, Deacon remarks that in sailing across the ocean to do business, the English beg God to look after them, which he does. But they are putting themselves "in the way of the devil," he claims, by trafficking in tobacco. "Let us not…for private gaine to our proper purses, procure the cause of a publick bane to our owne countrey people."[37]

Surprisingly, opposition to the cultivation of tobacco in Virginia also came from the very group who profited from its success: the Virginia Company itself. In 1620 a broadside was published in the form of a letter from the Company based in London to the leaders of the colony residing in Virginia. The broadside reports that "a great mortality due to disease" has happened in the colony and this is evidence that God was chastising them for misbehaving. The cause of God's displeasure was obvious, "the planting of tobacco and the neglect of other more solid commodities have not only redounded to the great disgrace of the countrey and detriment of the colony, but also in point of profit greatly deceive them which have trusted to it."[38] Various alternatives to tobacco were recommended that would benefit the colony and the English nation as a whole, including iron, cordage, pitch and tar, timber, silk, vines and salt. Provisions had been sent for the building of sawmills, as had experts in vine growing. None of these projects prospered, and despite opposition from England, tobacco remained Virginia's staple crop, thus defining the demographic development of the colony.

The lottery mentioned above, which was devised to raise awareness and money for the colony, was managed inefficiently and brought the Company's efforts into disrepute. Apparently many towns were slow to return their books, thus delaying the drawing of the lottery. In July 1615 in a letter to the city fathers of Exeter, Richard Martyn passed on the gratitude of the Virginia Company that Exeter had returned their lottery book completed and on time. In 1620 King James eventually became annoyed, because although funds were collected, the lotteries were hardly ever drawn. He rescinded the Company's right to hold the lottery, saying that it had caused "the hinderance of multitudes of our subjects." This debacle, caused by the disorganisation and corruption in the Company itself, must have angered many investors and ruined the reputation of the Virginia enterprise, leaving observers to think that this was just one more in a long list of schemes of unsuccessful and perhaps fraudulent schemes.

However, Virginia and her supporters were not the only targets who were lambasted by their opponents. New Englanders, too, were criticised from the earliest days of their emigration. There was a long tradition of ridiculing Puritans and their beliefs, typified by Shakespeare's portrayal of the humiliated Malvolio in "Twelfth Night." One of the most important mediums for satirising was the ballad, and this was the genre chosen by the anonymous author of "My brethren all attend me," a ballad poking fun at the Puritan migrants who went to Massachusetts in 1629-30. This ballad purportedly spoken by a Puritan trying to encourage his family to go to New England with him satirises the Puritan belief in the brotherhood of the chosen elect and criticises them for abandoning England. The chorus of the ballad is as follows:

> Come Along
> Leave this place of superstition
> Were it not for me ye Brethren be
> You'd sink into perdition. [39]

The inclusion of America as a topic in ballads and plays shows how significant news of these colonies was in English society in the early seventeenth century, and also shows how far this information had infiltrated the mindset of early modern Englishmen. Printed news from America continued to be of interest, but it had also become an integral part of fictional representations designed to entertain rather than inform. As an institution the theatre appealed to all classes and therefore if information about America was introduced into plays, then it reached all levels of society. Logically the audience must have understood the references to America in order to find amusing the subversive twisting of hope and plenty in the narratives. The most elite theatrical productions also made mention of England's colonies, without being overtly critical for fear of offending those watching the court masques who probably also invested in the colonial companies. George Chapman's masque of 1613, performed at Middle Temple, did parody the excessive ambitions of some of the adventurers to seek gold, by portraying Native Americans sitting on mountains of gold paying homage at Princess Elizabeth's wedding.[40] This portrayal came at a time when most people realised that the dreams of finding mineral wealth to rival that of the Spanish colonies were fruitless, as settler Ralph Hamor confirmed in his tract of 1615, saying that those in Virginia now hoped to find only the lesser minerals such as iron and alum.[41] So, Chapman's representation was not intended to be inspirational but rather parodied the failed ambitions of the colonists.

However, it was the popular London theatres that spread the word on the New World to a more diverse audience than any other medium apart from perhaps the broadsides. Plays performed in public in venues all over London and then in the provinces by travelling players, reached audiences that the printed medium could not achieve until the spread of literacy during the eighteenth century encouraged mass distribution of the newspaper and the magazine. Many of the Jacobean playwrights who fleetingly included America in their stories did not portray her favourably, or at least poked fun at those who went there. William Crashaw, the minister who spoke out in support of the Virginia enterprise was keenly aware that alongside the traditional enemies, the "papists and the devil," Virginia had a third category of opponent: the "players." Crashaw dismissed the slanderous words of the actors "they abuse Virginea but they are but players, they disgrace it true but they are but players."[42] Ben Jonson was probably the most famous satirical playwright of his day, so it is no surprise that he chose a subject as topical as the New World to target in the scenes of several of his plays. *Eastward Hoe*! was printed for the first time in 1605, but prior to that was performed in the Blackfriars Theatre. Like Chapman's masque, *Eastward Hoe*! parodied the assumption that gold would be found everywhere in North America. Jonson described Virginia as a country in which "all their dripping pans and chamber pottes are pure gold and…all the prisoners they take are fettered in Gould." He also closely mimicked the language of the promotional tracts, with a humorous on-stage delivery designed to cause hilarity although the words could have been lifted from a travel narrative. He said that the land was "temperate and full of all sorts of excellent viands." In an echo of Thomas More's *Utopia*, which would have been recognised by the more educated members of the audience, Jonson claimed that "you may be an alderman there and never be a scavenger, you may be a nobleman and never be a slave."[43] In another play, Jonson is even more blunt with his satire against America. *The Staple of News* depicted a group of English cooks migrating to America to try to convert the natives and make them "good eating Christians"![44] Thomas Middleton was another famous playwright of the early Stuart period who seemed to have nothing good to say about America. In his play of 1612, *No Wit, No Help like a Woman's*, the character Mistress Low-Water describes an eclipse in "Mexicana and California" but says that "we have no business there."[45]

Later on in the period, Phillip Massinger's *A City Madam*, performed from 1632 onwards, although not published until 1658, attacked the English men and women who chose to migrate to Virginia. He drew on standard criticism of early Virginia (and indeed the Caribbean Islands) that only the most unsavoury of men and women migrated there. One of the main characters, Luke, suggests sending his daughters to Virginia to marry planters and to help convert the natives. His suggestion is answered with howls of disapproval from the other characters on stage:

> *Lady Frugal*: How Virginia! High Heaven Forbid. Remember Sir I beseech you, what creatures are shipp'd thither.
> *Anne*: Condemned wretches, forfeited to the law.
> *Mary:* Strumpets and bawds, for the abomination of their life, spew'd out of their own country.[46]

Interestingly, Philip Massinger's interest in Virginia was aroused by reading of John Smith's *Map of Virginia* of 1612, one of the most popular of the travel narratives throughout the seventeenth century.[47] The travel narratives reinforced Massinger's fears about the dissolute nature of Virginia society. Although in the Civil War era men such as William Bullock (who incidentally never went to Virginia) claimed that the colony was a very prosperous and safe place for gentlemen to send their daughters, many other authors railed against the disorderly society there.[48] As early as 1609, William Crashaw complained that men had gone to Jamestown "for ease or idleness," and from many accounts, during the "starving time" law and order did break down completely.[49] As well as satirising the contemporary news from America, these playwrights were hostile to the colonial enterprise for another reason. A number of projects plagued seventeenth century society, and while not all these schemes were corrupt or unscrupulous many failed, despite proclaiming lofty ideals, leaving many commentators very cynical. These projects were promoted with similar language to that of colonial supporters, hence the association of America with other failed schemes. All claimed that they would provide the poor with work and the investor with profit.[50]

Part 3: Interpretation

It is impossible to take the early modern understanding of America in isolation from other cultural phenomena of the time. It is also not valid to separately interpret the promotional and critical literatures about the New World. These two strands of commentary were inextricably linked. An analysis of the way America was interpreted based on print culture, manuscript and word of mouth, and the cultural forms such as the theatre, is inevitably complex and multi-faceted. However, the emphasis within this relationship changed over time too. The ways in which the regions of North America were perceived were very different. This is not to return to a historical interpretation that sees Virginia or New England as the typically American example, the true forerunner of the United States. Rather, it is important to acknowledge that these regions were very different, both in their physical geographies, climate, flora and fauna, and also in the settlement that emerged. This meant that people who had never been to Virginia, New England or Newfoundland formed a distinct impression of each based on the literature written by those who had explored or settled those areas. It must also be remembered that the English Caribbean colonies created their own identity based on their own physical realities and economic development.

The reactions to American settlement found in print culture did not act independently from one another and they were not isolated from the world of oral communication. Especially in London and the port towns of the south coast, word of mouth was the most significant communicator of news about the New World during this period, as sailors and returning colonists told exciting and exotic tales. No doubt the most popular and lasting of these stories were those of extreme heroism in the face of adversity. A storyteller would know that his or her audience did not want to hear mundane accounts of the colony's survival. They wanted tales of extraordinary creatures, savage natives and intense weather. The interaction of these stories and those told in printed artefacts is very complex and it is almost impossible to ascertain which parts of an individual's knowledge came from oral and which from written sources. Certainly the authors of promotional literature were aware that the rumours spread by sailors and disgruntled company servants were very powerful in turning investors and potential migrants away from the colonial enterprise. The content of the promotional pamphlets was aimed at quashing the negative rumours that had begun to spread orally, thus positive and negative understandings of America interacted. Some of the over-enthusiastic early reports later caused a backlash when settlers found that America was not the land flowing with milk and honey that had been predicted. This readjustment led to a changed tone in the reports from Virginia from about 1612 onwards.

A further complication is that the opinions of individuals did not remain static, but rather evolved as the colonies themselves developed. The classic example is John Smith, probably the most famous and successful of the commentators on America. Smith's descriptions of Virginia mirror his own turbulent relationship with the Virginia Company. His first tract in 1608 is mostly positive about the opportunities in the new colony, but by 1612, when *A Map of Virginia* was published and Smith had already left Virginia in disgrace, he showed more awareness of the potential dangers in Virginia and especially of the power of the native tribes.[51] Finally, in 1622, Smith specifically addressed the weaknesses of the Virginia colony in his tract, *New Englands Trials*, laying the blame squarely at the feet of the Virginia Company. He wrote that "the honourable Company have been humble suiters to his Maiestie to get vagabonds and condemned men to go thither…so much scorned was the name of Virginia some did chuse to be hanged ere they would go thither."[52] This extract also shows the power of rumour in affecting the reputation of the colony, which after the massacre, was at its lowest point.

Throughout the period between 1607 and 1630, perceptions of Virginia ebbed and flowed. The two crisis points were 1610 and 1622: in 1610, when ships reached England bearing news of the horrific "starving time," the reputation of the colony was such that few people dared invest or travel there. The Company mobilised its preachers and authors to inspire the English to take heart and support the project again, arguing that the famine had been caused by errors in judgement of its leaders who had now been replaced. However, it was not until the tobacco boom of the mid 1610s that the good name of the colony was restored. The massacre of 1622 caused the reputation of Virginia in England to sink once more. The Company again employed its supporters to issue reassurances in print, but it took a long time for confidence to be restored. After 1630, although Virginia had achieved a modicum of economic stability, attention in England increasingly focused on new colonial ventures further north.

In conclusion then, what was the significance of print culture in creating an understanding of North America? By acting in concert with news spread orally and through manuscript distribution, print culture did affect the ways in which the English reader perceived America. In an era before printed newspapers, different artefacts such as the broadside and the pamphlet were used to spread information. Although literacy was still low amongst all but the elite of society, printed matter was consumed communally, giving illiterate individuals access to printed materials and to their revelations about the New World. While the contents of these books mirrored the events in the colonies, they also revealed a more complex intertextuality, reacting to other printed works to create an overall impression of the colonies. A single book often answered criticisms spread orally or by representations on the stage, showing the extent to which knowledge on America was entrenched in English culture.

What was the significance of these representations? They give the historian an access point into the *mentalité* of seventeenth century England, not just at the level of James' court, but throughout society. These printed artefacts also influenced the development of the colonies across the Atlantic. Although it is often imagined that the communication between the two hemispheres was limited, people carrying news regularly travelled across the Atlantic during this early colonial period. The way that England perceived the colonies was always incredibly important to the settlers. Even though they were far away from home, they maintained that they were still Englishmen. They were keenly aware of criticisms from England and they did their best to contradict rumours that early Virginia was full of idlers and criminals. The colonists also put pen to paper in order to defend their reputations, thus contributing to the growing printed knowledge about their new home. There were exceptions to this rule but, for the most part, the flow of interpretations and representations between London and the colonies was truly two-way.

---

[1] R. Hakluyt, *The Principall Navigations, Voyages and Discoveries of the English Nation* (London, 1589); S. Purchas, *Purchas His Pilgrims*, XX vols, (Glasgow, 1957). See D. Armitage, *Ideological Origins of the British Empire* (Cambridge, 2000), 68-85.

[2] See W. Sherman, "Stirrings and Searchings," in P. Hulme & T. Youngs, *The Cambridge Companion to Travel Writing* (Cambridge, 2002).

[3] The premise that print culture had a gradual not a revolutionary impact on changing the English epistemology is derived from E. Eisenstein, *The Printing Press as an Agent of Change* (Cambridge, 1979), 227.

[4] T. Watt, *Cheap Print and Popular Piety 1550-1640* (Cambridge, 1991). This concurs with Louis B. Wright's earlier assessment in *Middle Class Culture in Elizabethan England* (Chapel Hill, 1935), 508.

[5] The authorities on this are still H.G. Aldis, *The Book Trade 1557-1625* (Cambridge, 1909) and H.S.Bennett, *English Books and Their Readers 1558-1603* (Cambridge, 1965).

[6] J. Brereton, *A Brief and True Relation of the Discovery of the North Part of Virginia* (London, 1602), 11.

[7] ibid, 18-20.

[8] C.Levett, *A Voyage into New England Begin on 1623 and Ended in 1624* (London, 1628), 1.

[9] J. Smith, *A True Relation of Such Occurrences and Accidents of Note...*(London, 1608), 35.

[10] J. Frederick Fausz, "Merging and Emerging Worlds," in L. Carr, P. Morgan & J. Russo, eds., *Colonial Chesapeake Society* (Chapel Hill, 1988), 51.
[11] *For the Plantation in Virginia or Nova Britannia* (London, 1609).
[12] *The Inconveniences that have happened to some of the persons* (London, 1622)
[13] 1613 broadside facsimile, in *Three Proclamations Concerning the Lottery for Virginia 1613-1621* (Rhode Island, 1947).
[14] *Londons Lottery*, part one (London, 1612)
[15] ibid, part two.
[16] A. Fitzmaurice, *Humanism and America: An Intellectual History of English Colonisation 1500-1625* (Cambridge, 2003); T. Scanlon, *Colonial Writing and the New World 1583-1671: Allegories of Desire* (Cambridge, 1999).
[17] W. Crashaw, *A Sermon Preached in London before the right honourable Lord Lawarre* (London, 1609), 32. William Axtell discusses this misunderstanding in his book *The Invasion Within: The Context of Cultures in Colonial North America* (Oxford, 1985), 131. Joyce Chaplin also explores this in *Subject Matter: Technology, the Body and Science on the Anglo-American Frontier* (Cambridge, Mass., 2001).
[18] W. Symonds, *A Sermon Preached at Whitechapel…* (London, 1609),57.
[19] P. Copland, *Virginias God Be Thanked* (London, 1622), 9.
[20] J. Donne, A Sermon upon the VIII verse of the 1 Chapter of the Acts of the Apostles (London, 1622) 13.
[21] Ibid, 34-35.
[22] *To the Virginian Voyage*, stanza 2.
[23] Ibid, stanza 11.
[24] C. Brooke, *A Poem on the Late Massacre in Virginia* (London, 1622), sig. C.
[25] J. Hagthorpe, *Englands Exchequer* (London, 1625), 32-33.
[26] R. Eburne, *Plaine Pathway to Plantations* (London, 1624), 125.
[27] Ibid, 104.
[28] J. White, *The Planters Plea or the Grounds of Plantations Examined and usuall Objections Answered* (London, 1630), sig. A2. See D. Underdown, *Fire from Heaven*, (London, 1992), 24-25.
[29] Ibid, 25, 29.
[30] S. Purchas, "Virginias Verger," in S. Purchas, ed. *Purchas, His Pilgrims*, vol. XIX (Glasgow, 1951).
[31] Letter from Lord de la Ware to the Earl of Salisbury, *Virginia Historical Magazine*, vol. 14, (1906-7), 381.
[32] S. Purchas "Account of Proceedings at Virginia," in S. Purchas ed. *Purchas His Pilgrims*, vol. XVIII, 538.
[33] R. Eburne, *Plaine Pathway to Plantations*, 38.
[34] J. Deacon, *Tobacco Tortured: or the Filthie Fume of Tobacco Refined* (London, 1616), title page.
[35] Ibid, sig. A1.
[36] Ibid, 82.
[37] Ibid, 182-3.
[38] *The Treasurer, Councell & Co for Virginia Writing to the Governor and Councell there Residing* (London, 1620)

[39] Bodleian Library, Ashmole MSS 36, f. 100.
[40] R. Cawley, *The Voyagers and Elizabethan Drama* (Boston, 1938), 337-338. For a more recent assessment, see T. Marshall, *Theatre and Empire: Great Britain on the London Stages Under James VI and I* (Manchester, 2000).
[41] R. Hamor, *A True Discourse of the Present Estate of Virginia* (London, 1615), 34.
[42] W. Crashaw, *A Sermon Preached in London before the Right Honourable Lord Lawarre* (London, 1609), 57.
[43] C. Hertford & P. Simpson, eds. *Ben Jonson* (Oxford, 1954), vol. IV, 569.
[44] R. Cawley, *The Voyages and Elizabethan Drama*, 301.
[45] T. Middleton, *No Wit, No Help like a Woman's*, in *A Mad World, Masters and other Plays*, ed. M. Taylor (Oxford, 1995), 251.
[46] P. Massinger, *The City Madam*, ed. T Craik (London, 1964), 76.
[47] Ibid, editor's notes.
[48] W. Bullock, *Virginia Impartially Examined* (London, 1649), 54
[49] W. Crashaw, *A Sermon Preached in London,* 41, 49.
[50] J. Thirsk, *Economic Policy and Projects* (Oxford, 1978).
[51] See, for example, J. Edwards, "Between 'Plain Wilderness' and 'Goodly Corn Fields'- representing land use in Early Virginia," in R. Appelbaum & J.W.Sweet, eds., *Envisioning an English Empire* (Philadephia, 2005), 217.
[52] J. Smith, *New Englands Trials* (London, 1622), 16.

CHAPTER TWO

AN EMPIRE RESTORED: AMERICA AND THE ROYAL SOCIETY OF LONDON IN THE RESTORATION

SARAH IRVING

Over thirty years ago, J. H Elliott framed the confrontation between the Old World and the New in terms of the issue of how the discovery of the New World affected European categories of understanding. America, he argued, was assimilated into "the half-light of [the European] traditional mental world."[1] More recently, however, some scholars have suggested a very different situation, in which the discovery of America produced a kind of shock of the new; a sense of wonder which, defying categorization, compelled the reorientation of European consciousness.

The epistemological nature of this issue has led historians to emphasise the importance of the New World as a storehouse of new knowledge, and to examine the impact of this knowledge on the intellectual pursuits of the Old World. Natural Philosophy, for example, has been the subject of some excellent studies of the role of knowledge in shaping British attitudes to the New World and to her burgeoning empire.[2] This chapter aims to extend this research by turning our attention towards the concept of empire. I want to suggest that, if we introduce into this line of enquiry an exploration of the language of empire, we discover that America was part of a largely neglected intellectual tradition.

We could be forgiven for assuming that the idea of empire in seventeenth-century England was that which is familiar to us today; a territorial, colonial exercise of a state's power. One of the sources which shaped the mental universe of early modern England, however, tells a markedly different story. The Creation narrative in the book of Genesis describes an empire which is not geographic. "Have dominion over the fish of the sea, over the birds of the air, and over every living thing that moves on the earth", commanded God to Adam in the Garden of Eden.[3] The dominion of which God spoke consisted in Adam's perfect knowledge of the natural world; his ability, as Genesis 1:20 put it, to give the proper names "to all cattle, to the birds of the air, and to every beast of the field." Adam's dominion was not a sovereignty of territory but a sovereignty of knowledge. In the Fall from Eden, however, Adam lost his omniscience, and with it his epistemic empire over the world. It is important to note here that in the seventeenth-century, the terms "dominion" and "empire" both denoted plenary authority or sovereignty, whether over territory or over intangible phenomena such as ideas.

The mid seventeenth-century was a transformative moment in the history of science, in which a number of natural philosophers reframed the Adamic story in an apocalyptic context. In the final ages of the earth, "knowledge will increase" (as the book of Daniel put it) and man would be restored to his empire over nature.[4] Natural philosophers, following the work of Francis Bacon, believed that it was their task to restore man's empire of knowledge. The Royal Society of London, England's first scientific institution which was established in 1660, set itself this project.

This chapter explores the way in which the Royal Society imagined America during the Restoration. The drive to restore perfect knowledge of the natural world motivated an extensive correspondence between fellows of the Royal Society of London, and colonists, traders and explorers in the Americas. The significance of this correspondence was not only, as many historians have recognised, that America was important as the setting for posing new questions about the veracity of knowledge, but also that America was *conceptually* significant, as it gave shape to a tradition of empire as dominion over knowledge.

I: Intellectual contexts

In October 1667, Henry Oldenburg, the Secretary of the Royal Society of London, wrote to John Winthrop Jr, the Governor of Connecticut. Oldenburg's aim was to remind Winthrop that his role was not only to effect good government in His Majesty's plantations; as a member of the Royal Society, Winthrop's responsibility was also to help foster what Oldenburg and his colleagues termed, a "Commonwealth of Learning."

> So good an opportunity as this I could not let passe without putting you in mind of yr being a Member of ye Royall Society, though you are in New England; and even at so great a distance, you may doe that Illustrious Company great Service…[by] communicating to them all the Observables of both Nature an Art, yt occur in the place, you are.[5]

Oldenburg's request for information about the natural world of New England alerts us to the importance of America in a project of "*lay[ing] open…an*

Empire *of Learning"* as the astronomer Edmond Halley put it.[6] In the forty years following the Restoration, Winthrop was just one of many correspondents of the Royal Society in the New World. Sometimes colonists who were Fellows themselves would send back "rarities," "curiosities" and detailed knowledge from the colonial periphery to the metropole of London. This transfer of knowledge was literary, but often tangible and haphazard too; letters and wooden boxes of specimens were shipped across the Atlantic, occasionally falling victim to the perils of sea voyages.

In order to understand the Royal Society's vision of a New World empire of knowledge we must firstly sketch the intellectual context in which it emerged. The story of the appearance of America in early modern natural philosophy is a story of a confrontation between ancient authorities and a strange landscape which challenged those authorities by virtue of its novelty. The ancients had not imagined America. It was a world unknown to Pliny and his Roman contemporaries, and unaccounted for by the Bible. This fact meant that, whilst America was interpreted within the framework of existing intellectual authorities—chiefly the Bible, Roman writers and the scholastic tradition—it also challenged their heuristic powers.[7] As Francis Bacon pointedly remarked of the Greeks in the *New Organon* (1620), "They knew only a fraction of the parts and regions of the world...much less the territories of the New World...But in our time large parts of the New World and the farthest parts of the Old are becoming known everywhere, and the store of experiences has grown immeasurably. Hence if (like astrologers) we are to gather signs from the time of nativity or conception, nothing significant seems to be forecast for those philosophies."[8]

Bacon's confidence in the novelty of the discoveries of his age belies the fact that one aspect of his intellectual project—the ideal of recovering the epistemic dominion over the earth that mankind once possessed—was not new. It is important to note that there was a long historical continuity of the idea of recovering knowledge of the world. As Richard Yeo has pointed out, "in Western tradition there has been a conviction that it is possible and worthwhile to collate knowledge that is representative of some larger whole...a view of the world as a mirror of the divine mind."[9] The idea that Adam's knowledge had been lost in the Fall but that elements had survived through the patriarchs, and thence through early theologians, was commonplace in Scholasticism as well as in medieval occult philosophy. The Hermetic tradition, for example, held that this original knowledge—the *prisca theologia*—had been passed down through Moses to the Egyptian priest Hermes Trismegistus and thus to Plato.[10] A related belief, held by the Cambridge Platonist Ralph Cudworth, was that elements of the true religion were to be found scattered and fragmented throughout all the world's religions.[11]

It would be tempting to see an over-arching concern with the restoration of man's original knowledge as a kind of early modern *zeitgeist*, because elements of this ideal can be traced in the work of thinkers as diverse as the Elizabethan magus John Dee, the radical Protestant Samuel Hartlib and the natural philosophers Francis Bacon and Robert Boyle. We must be cautioned, however, against such an assumption. This particular episode of intellectual history was more complex. Sixteenth and seventeenth-century English natural philosophy was animated by a diversity of intellectual traditions which, while sharing a broad interest in the re-creation or emulation of paradise on earth, were markedly different in their authors' intentions and the sources from which they drew their inspiration. Recognising this complexity, we can identify a number of possible ways of imagining America which were available to early modern English writers. It will be helpful here to outline two, in order to understand the context in which the idea of America as a restored Adamic empire of knowledge emerged.

Perhaps the most significant way of conceiving of America was through a set of questions about the origins of the New World's inhabitants. How does one make sense of the Americans in terms of the descendants of Noah's sons? Did the Biblical Flood reach America? Are the Americans one of the lost tribes of Israel? And above all, why have they not established commonwealths? Such questions prompted reflections upon this new land and its strange inhabitants. Here the English were mirroring questions that the Spanish had also asked when they first encountered America. Columbus and then the conquistadors had been particularly interested in how the Amerindians could be fitted into the Biblical history. One particular concern, as Sabine McCormack has shown, was with the similarities between the myths and deities of the Amerindians, with those of the Bible. The Salamanca theologian Bartolome de las Casas, for example, maintained that the Inca worship of the sun god "augured the worship of Christ the Sun of Justice in the Andes."[12] The similarity between the creation-narratives of the Americans with those of the Bible convinced the Calvinist missionary Jean de Lery that the Tupinamba people of Brazil's myth of a flood "contained traces of the story of Noah."[13]

In England, the Norfolk minister Thomas Thorowgood argued in *Jews in America or The Probabilities that the Americans be of that Race* (1650), that the American Indians were in fact Hebrew, being descended from Noah's son Shem. The ancient Hebrews were not the only race with which the Americans were compared. Johann Theodore de Bry who illustrated the mathematician and explorer Thomas Hariot's account of Virginia in 1590, depicted the ancient Picts as Americans. As Amy Gordon and Peter Burke have both argued, these kind of comparisons helped in part to generate a new mode of historical writing which concerned itself with the cultural and social origins of peoples, rather than with their politics.[14] Of course the origins of anthropology, as many scholars have rightly observed, were not far off.[15]

A number of English texts, including Thorowgood's, grappled with the issue of the Noachian flood. The issue of whether or not the flood had reached the New World, and if so, from which of Noah's sons its inhabitants were descended, was vexing. In 1681 Thomas Burnet, who later became the Royal Chaplain to William III, published *Telluris Theoria Sacra, (The Sacred Theory of the Earth.)* Strongly influenced by Descartes' *Principia Philosophicae*, Burnet argued that the current state of the earth, with its oceans, valleys and mountains, was evidence of the imperfections caused by the great deluge. Burnet then tried to explain the flood by arguing that when the world was created, the water retreated to the core of the globe, becoming artesian water which rose to the surface when God commanded it.[16]

The set of questions about the origins of the people and the natural environment of the Americas provided one possible framework for interpreting the New World. Closely related to this set of questions, and very often drawing upon it, was a second context in which America was imagined. This was the intellectual tradition of natural history writing and the practice of collecting objects with which it was increasingly associated.

An interest in natural history had existed since ancient times but was transformed considerably in the late Renaissance and early seventeenth-century. The Roman author Pliny the Elder's *Naturalis Historia* (c.77AD) was one of the most frequently cited texts by early modern natural historians. Pliny's authority, for example, is cited in the Jesuit missionary Jose d'Acosta's *Historia natural y moral de las Indias* (1590) and by the Spanish historian Gonzalo Fernandez de Oviedo yValdés, *General and Natural History of the West Indies,* the English version of which was published in 1555.

The medieval interest in the natural world derived from the belief that nature held allegorical keys to understanding God's two books. One of the defining characteristics of medieval natural histories was their concern with the heuristic significances of animals. As William B. Ashworth Jr points out, Conrad Gesner's *Historia Animalium* (1555) for example, included folk tales and myths about animals.[17] Herbs as well as animals were of particular interest in the Renaissance, which is evidenced by the genre of herb collections called the *res herbaria.* As Scott Atran has shown, however, the interest in the variety of herbs did not necessarily mean that an attempt was made to cite the exact location of a particular herb, or its variety. In fact, Scholastic writers resorted to ancient categories for describing herbs, even when the herbs came from the New World.[18]

This reliance upon ancient authorities for the interpretation of the natural world was accompanied by a view of the significance of nature as emblematic, laden with symbols of the divine. Scholars frequently argue that this was a distinctly "pre-modern" conception of nature because its heuristic principle was that of symbolic reference. It was then eclipsed, they believe, by the taxonomic description characteristic of modern science.[19] Michel Foucault memorably described this change as an epistemic shift.[20] As Peter Harrison has recently shown, the contrast between interpreting texts and nature allegorically, and reading them as literal truth, was the contrast between the Catholic and Protestant hermeneutics.[21] To simplify Harrison's point, Catholics read the Bible and nature allegorically, whereas Protestants read the two books as literal history. This new hermeneutic was central, he argues, to the rise of modern science.

This epistemological transformation is perhaps easy to overstate, but there were some important associated changes in the practice of natural history and collecting in early seventeenth-century England. The first was an increasing interest in collecting natural specimens which was made possible by explorations of America. This phenomenon helped give rise to the idea that reliable knowledge about nature was that which was directly discovered rather than that which was derived from ancient texts.

During the middle ages and early Renaissance, collections of natural and artificial objects, just like natural histories, were seen as a representation of God's power to intervene in natural processes and produce miracles.[22] As Lorraine Daston has pointed out, travel "was the alpha and omega of collecting…the voyages of exploration and subsequent trade with newly discovered lands created a steady flow of exotica."[23] A good illustration of the increasingly important role that travel played in collecting is the case of the John Tradescants, senior and junior. John Tradescant senior was the gardener to Charles I, and amassed a great collection of plants. Although he possessed American specimens, he had not seen it necessary to travel to the New World himself. By contrast, several decades later, his son John Tradescant the younger (c.1608-1662), whose collections formed the basis of the Ashmolean Museum, realized the importance of the New World to natural knowledge. Tradescant Junior took an avid interest in the Americas. He visited Virginia three times, first in 1637, then in 1642 and finally during 1653-4. The Calendar of State Papers records that "In 1637 John Tradescant was in the [Virginia] colony, to gather all rarities of flowers, plants, shells &c."[24] America was fast being imagined as an Eden; a storehouse of the unknown natural world ready for discovery and classification.

In 1656, Tradescant Jr published a catalogue of his collection, entitled *Musaeum Tradescantium.* Funded by Elias Ashmole, it was the first catalogue of a museum printed in England, but its categories and classifications were only a semblance of order.[25] The culture of collecting in mid seventeenth-century

England remained largely in the form of cabinets of curiosities, categorized not by type or geographic location but often by the material from which it was made, if at all. Objects hung from ceilings and the purpose of most collections was not to improve knowledge of the natural world, but to represent social status through the conspicuous display of rare objects. Francis Bacon's proposed reform of natural history was intimately connected to the disdain he felt for the way in which collecting was being practiced in the Elizabethan and Jacobean courts, which he saw one of the weaknesses of the existing state of knowledge. The character of natural history in late sixteenth and early seventeenth-century England was unsystematic and amateur. Collecting specimens of natural history was a popular pastime of the aristocracy, but Bacon viewed such collections as focusing only upon the curiosity of the artefact; they were aimless, the objects were de-contextualised, and the trivial collections did not advance knowledge of their subject. Meticulous and empirical natural histories, Bacon argued, would be one of the most fruitful ways of advancing knowledge. The epistemological concern underlying Bacon's project for natural history was with the purpose and veracity of knowledge. It was precisely this problem, however, which haunted the Royal Society's attempts to regain man's original omniscience.

II: Eden and the New World

This brief sketch of two of the available ways of imagining the New World provides a framework for understanding the Royal Society's idea of America as central to the recovery of man's Edenic empire. As John Prest has shown, there had been for centuries a conviction that the Garden of Eden was still to be found on earth. It was just a matter of finding it. In *The Divine Comedy,* for example, Dante put forward the idea that the Garden of Eden lay directly opposite Jerusalem, but in the southern hemisphere.[26] In the Medieval period Sir John Mandeville claimed to know that the Garden existed, despite not having been there himself, and cartographers frequently depicted the Garden on their maps in places as yet undiscovered, such as in the kingdom of Prester John, or in Africa or India."[27]

Many English accounts of the New World by sixteenth-century travelers were animated by the belief that, if the New World was not the Garden of Eden, it was certainly part of the world described by the Old Testament. In his *Briefe and True Report of the New Found Land of Virginia,* (1588), for example, Thomas Hariot described Virginia in such terms. Richard Hakluyt believed that Eden would be discovered in the southern hemisphere.[28] The imagination of the New World as Eden had important consequences for the practice of legitimating English colonial expansion. Richard Drayton has shown that the Garden of Eden embodied the idea of redemption through the enclosure and cultivation of the

New World. There was a connection between using agriculture to return the world to its Edenic state—making it fruitful once more—and the ideological justification for colonisation.[29] In a similar vein, Patricia Seed argues that it was through planting gardens that England ceremonially enacted its dominion over the Americas.[30] The idea of Eden, therefore, enabled the English to imagine America as an uncultivated land which they could possess by re-enacting Adam's first labour. As John Locke stated in a phrase which, whatever his original intention, was later used as one of the most important legitimations of British imperialism: "Whatsoever then he removes out of the State that Nature hath provided, and left it in, he hath mixed his *Labour* with, and joined to it something that is his own, and thereby makes it his *Property*."[31]

This scholarship has provided excellent insights into the importance of the idea of Eden as a framework through which the New World was imagined, but the significance of the idea of an Edenic America to the *conceptual* history of empire has not been explored. We can build upon the existing scholarship, I think, by posing a new historical question. How did the idea of man's original, pre-lapsarian knowledge contribute to the seventeenth-century development of the *concept* of empire?

I would argue that the Fellows of the Royal Society made two intellectual manoeuvres with the idea of the Garden of Eden. The first was to state its epistemological significance as the site of man's original omniscience, and the second was to conceptualise this omniscience as an idea of *empire*—as a theory of sovereignty—and argue for its restoration through the proper pursuit of natural philosophy in America.

The Royal Society certainly conceived of the New World in Edenic terms. In the 1674 edition of the Royal Society's journal, the *Philosophical Transactions,* its editor Henry Oldenburg reported that

> The very Wilderness of New England, is, on a sudden, become a fruitful Orchard, fenc'd with Ships of their own building...And in less time, by God's Blessing, Jamaica may be the fairest Garden of the World.[32]

The importance of the Edenic state of the Americas was, however, not only its fruitfulness. It was also the fact that it contained the knowledge about the natural world which had been lost in the Fall. The initial ideological manoeuvre the Fellows of the Royal Society made was to state the epistemological—rather than simply theological—significance of discovering Eden.

Robert Boyle explained in his essay "On Final Causes" that,

> Tis recorded in the Book of Genesis, the Design of God in making man, was, that men should Subdue the Earth (as vast a Globe as 'tis) and have dominion over the Fish of the Sea, and over the Fowle of the Air, and over the Cattle, and over all the Earth and...over every living thing that moveth upon the earth.[33]

The basis of man's primordial empire was knowledge: "Man's power over the Creatures depends chiefly upon his knowledge of them."[34] It is little surprise, then, that Boyle conceived of the project of Natural Philosophy as recovering "a lost empire over the works of nature."[35]

Here Boyle makes a second ideological manoeuvre, conceptualising man's original knowledge of the world as an empire. This idea was commonly held. There is not sufficient space to cite more than one example here, but Joseph Glanvill, the Church of England clergyman, echoed Boyle's emphasis upon the epistemological powers of Adam and his descendents when he argued in his essay "The Usefulness of Real Philosophy to Religion," that the patriarchs "were well instructed in the *Knowledge* of *God's Works,* and contributed to the good of Men by their useful Discoveries and Inventions. *Adam* was acquainted with the *Nature* of the *Creatures.*"[36] Recovering this encyclopaedic empire of knowledge of all the creatures required knowledge gleaned in the New World:

> That so many and so various creatures as we have mentioned, to be directly subjected by God to man's dominion, may be of great advantage to him, and of more than they have been, when once they shall be improved by sagacious and industrious virtuosi, may appear very probably, if we consider what great benefits accrue not only to single persons, but to whole communities, and sometimes even to nations by two or three vegetables, as many reptiles, and as few minerals, of most of which the uses were unknown to the ancients, and those of the rest but little known… The other plant, though reckoned but a week, makes at this day a great part of the commerce between *Europe* and some American country, especially *Virginia*, the latter of which country sends yearly to England alone a considerable fleet freighted almost only with tobacco…All which improvements ought to excite man's gratitude to Him, that made those creatures to his hand, and endowed him with a rational faculty and fit organs to exercise his plenary dominion over them.[37]

III: Natural History and the New World

> Here Botaniques are discriminated, ranged and explicated by the most Essential indications of Nature….New England and the other English Plantations in America acquaint us with their Vegetables and other rarities."[38]

As Henry Oldenburg made clear in his dedication of the *Philosophical Transactions* of 1672, man's epistemic empire would be recovered with knowledge gleaned from the New World. The best point of entry into understanding how the Royal Society imagined America as part of an empire of knowledge is through an examination of two different practices through which they set about accumulating knowledge from America. The first was the attempt to compile an encyclopaedic natural history of its plants, animals, climate and soil. As Oldenburg explained to his correspondent Rene Francois Sluse, "we seek thoroughly to scrutinize everything—the heavens, the Earth, the subterranean world; the air, the meteors, and stars; rivers, seas, vegetables, minerals, animals; so that there is nothing of all this that you may not explore, wherever you turn."[39]

The sheer encyclopedic nature of this natural history meant that the Royal Society relied heavily upon the Americas to provide the necessary information. Oldenburg's letter to Charles Hotham, a Fellow of Peterhouse, Cambridge and a non-conformist minister from Wigan who moved to Bermuda in 1662, gives a nice illustration of the specificity and detail of American requests:

> 1.What is ye variation of ye Needle at the Bermudas?...5. To inquire after the natural productions of that Iland of ye Baham's, wch is call'd New Providence, where abundance of considerable things are said to be found, as to Plants, Trees, Fish and Fowl. 6. To send over some specimens of ye roots, seeds and Fruits of the Bermudas. 7, What is conceived to be ye cause of the Longevity of ye inhabitants of ye Bermudas, and the Bahama Ilands.[40]

A close reading of the *Philosophical Transactions* reveals that in almost every edition of the journal between its first publication 1665 and 1700, there was at least one report pertaining to the natural environment of the New World. The format and style of these reports varied. The most common type were copies of letters sent to the Society from correspondents in the Americas. John Clayton, the Rector of Croston at Wakefield, for example, sent a letter "giving a farther account of the Soil, and other Observables of Virginia."[41] In 1695, Sir William Beeston, the Governor of Jamaica, had a letter published which contained his observations about the barometer and of a hot bath in Jamaica.[42] John Winthrop wrote back to the Royal Society many times, describing natural artifacts of New England in detail. Speaking of his recently acquired knowledge, Winthrop reported, "I know now, whether I may recommend some of the productions of this Wilderness as rarities or novelties…There are…small Oaks, which though so slender and low…have yet Acorns and cups upon them".[43]

A second type of report regarding the New World was the letters from Fellows within England recounting information that had been sent to them. When John Locke, in his capacity as Secretary to the Board of Trade, came across information about the natural history of the Americas, he copied it out and forwarded it to Oldenburg, who increasingly became a conduit for such information. As Secretary of the Royal Society, Oldenburg would often choose to publish extracts from his American correspondence in the *Philosophical Transactions*. In May 1673, for example, Locke sent Oldenburg "an account I lately received from New: Providence one of the Bahama Islands concerning fish there."[44]

Another Fellow of the Royal Society, the physician and naturalist Martin Lister, also sent information he received to Oldenburg. Lister's correspondent was one Mr. Thomas Townes, who was born in Barbados and returned there after matriculating at Christ's College, Cambridge in 1664. In 1673, Lister wrote to Oldenburg and transcribed "an observation or two" which Townes had sent him.

> I have heard it questioned, whether America have not some plants common wth those of Europe, especially ye more Northern parts of it. The soil here is fertile, though not above a foot or two thick upon a whole & spongie lime-stone rock, whch affords good Quarries here & there…Indeed ye whole Island appeares in a manner like a scattered town, wch wth ye perpetual green fields & woods makes this place very pleasant.[45]

A third type of entry in the *Transactions* which dealt with America was the book review. In 1672, for example, there is a review of a book entitled, *The American Physitian; or a Treatise of Roots, Plants, Trees, Shrubs, Fruit, Herbs &c growing in the English Plantations in America.* The reviewer says that the book is "of good use, forasmuch as it may make a part of the Universal History of Nature."[46]

Such a universal history was in part required because of a widely held dissatisfaction with the existing state of knowledge of plants. It was this sentiment which sparked Sir Hans Sloane's interest in the plants of Jamaica and the Caribbean. The physician, naturalist and later the founder of the British Museum, Sloane recounted that the information about plants "which I met with in Collections were common in the West-Indies, …were not so Satisfactory as I desired." This, he continued, "incited me to do what I could to be no useless Member [of the Royal Society] but to cast my Mite towards the Advancement of Natural Knowledge, and the Faculty of Physic."[47]

In 1687, Christopher Monck, the Duke of Albermarle and Governor of Jamaica, offered Sloane the position of his personal physician. Sloane set sail to the New World via Madeira and the Canary Islands. The party arrived in Jamaica in December that year, and they remained there until October 1688 when the Duke died. Sloane departed for England in March 1689, carrying with him approximately 800 plant specimens. The resulting natural history, published in 1696, became famous. The work was titled *Catalogus plantarum quae in insula Jamaica sponte proveniunt*, and was dedicated to both the Royal Society and the Royal College of Physicians. Sloane also published a more general work of natural history about his voyage, entitled *Voyage to the Islands of Madera, Barbados, Nieves, S Christophers and Jamaica, with the natural history of the last of those islands; to which is prefixed an introduction, wherein is an account of the inhabitants, air, waters, diseases, trade &c.* When Sloane returned to England he showed his eight-hundred plant specimens "very freely to all lovers of such Curiosities."[48] Sloane's collection, which by his death included 71,000 objects, a library and herbarium, became the foundation for the British Museum which was first opened to the public in 1759.

Royal Society Fellows who could not travel to the Americas played their part in trying to collect an encyclopaedic knowledge about them through correspondence instead. Oldenburg, for example, never traveled beyond Europe, but his correspondence served the same purpose. He implored Winthrop to compose "a good History of New England, from the beginning of ye English arrival there, to this very time, containing ye Geography, Natural Productions, and Civill Administration thereof, together wth the Notable progresse of the Plantation, and the remarkable occurrences in the same."[49]

The detail and categorization of the information which Oldenburg sought illuminates one of the defining characteristics of the way that the Royal Society set about trying to create an encyclopaedic natural history. This was its attempt to be meticulously systematic. The injunction to organization in the pursuit of natural history was a legacy of Bacon, who made it clear in the *Parasceve* that a series of questions were the best methodology.[50] In the eighth issue of the *Transactions,* published in 1666, the astronomer Laurence Rooke devised a set of "Directions for Sea-men, bound for far Voyages."[51] They included, in part:

> 3. To remark carefully the Ebbings and Flowings of the Sea, in as many places as they can, together with all the Accidents, Ordinary and Extraordinary of the Tides; as their precise time of Ebbing and Flowing in Rivers, at *Promontories* or *Capes;* which way their current runs, what Perpendicular distance there is between the highest Tide and lowest Ebb, during the Spring Tides and Neap-Tides; what day of the *Moons* age, and at what times of the year, the highest and lowest Tides fall out: And all other considerable Accidents, they can be observe in the Tides, chiefly near Ports, and about Ilands, as in St. *Helena's* Iland, and the three rivers there, at the *Bermudas* &c.

In addition to these detailed instructions, five months later Robert Boyle set out the *General Heads for a* Natural History of a Countrey, *Great or Small*"[52] in the *Philosophical Transactions.* "*First,* in the Earth *it self,* may be observ'd, its dimensions, situation, East, West, North, and South: its Figure, its Plains, and Valleys, and their Extent; its Hills and Mountains, and the height of the tallest, both in reference to the neighbouring Valleys or Plains, and in reference to the Level of the Sea…Whether the Countrey be coherent, or much broken into Ilands. .. what the nature of the Soyle is, whether Clays, Sandy…As also, by what particular Arts and industries the Inhabitants improve the Advantages, and remedy the Inconveniences of the Soyl…*Secondly,* above the ignobler *Productions* of the Earth, there must be a careful account given of the

Inhabitants themselves, both *Natives* and *Strangers,* that have been long settled there: And in particular, their Stature, Shape, Colour, Features, Strength, Agility, Beauty, (or the want of it) Complexions, Hair, Dyet, Inclinations, and Customs that seem not due to Education."[53] Boyle's detailed instructions continued for several pages.

In the next volume of the *Transactions,* inquiries were published relating specifically to different countries. "*Inquiries For* Virginia *and the* Bermudas," for example, read:

> 1. Concerning the Varieties of Earths; 'tis said, there is one kind of a *Gummy* consistence, white and cleer: Another, white, and so light, that it swims upon water: Another, red, call'd Wapergh, like *Terra Sigillata.* Quaere, what other considerable kinds are there? And to send over a parcel of each. 2. What considerable Minerals, Stones, Bitumens, Tinctures, Drugs? 3. What hot Baths, and of what Medicinal use? 4. What is the Original of those large Navigable Rivers, which empty themselves into the Bay of *Chesapeak?* And whether on the other side of that ridge of Mountains, from which they are supposed to proveed, there be not other Rivers, that flow into the *South-Sea?...*13. Whether round about the Coast of the *Bermudas,* the *Tydes* keep the same time; and at what a clock precisely 'tis *High-water* on the Dayes of *Full* and *New* Moon; and how high the Water rises then? And the like on the Coasts of *Virginia* and *Florida?*[54]

The next series were directed at "Guiana and Brasil" and included questions about "whether upon the the Leaves of that *Brasilian* Tree, call'd *Cereiba,* there is, in a Sun-shiny day, found a *White Salt,* in that quantity, that one may gather as much from two or three Leaves, as will well salt a good pot of Broth?" There were so many inquiries, that

> The other Inquiries, ready for the other Countries above-named, are, to avoid tediousness, referred to another opportunity."[55]

The following year, the introduction to the *Transactions* proudly proclaimed that "even our *former Tracts* … have already brought in several pertinent *Answers; viz* from a *Sea-Voyage,* the *Caribbe*-Islands, and *Jamaica.*"[56] In that edition of the journal, there followed "*Enquiries and Directions For the* Ant-*Iles*, or Caribbe-*Islands*".[57] The section "Of Vegetables", for example, included the question "whether the Juice of the Fruit of the Tree *Junipa,* being as clear as any Rock-water yields a brown Violet-dye?"[58] The queries are too numerous to quote at length, suffice to say that for the Antilles and Caribbean, they continue for six pages.[59]

It is important to note that although Oldenburg's *Transactions* makes the aims and processes of the Royal Society's imperial epistemological project seem straightforward and even well-organised, the reality was rather different and more complex. For one thing, the success of the Society's inquiries about natural knowledge was open to debate. During the first three decades of its existence, the Royal Society received a fair amount of criticism, and even mocking, for its iconoclastic adventures in the reformation of science.

One of the Society's most vehement critics, the physician Henry Stubbe, believed that the Society knew far less about the New World than it purported to. "Just so when I went to Iamaica and desired that Honourable Personage Mr. Robert Boyle to procure some directions for Philosophical Inquiries in that Countrey; He, with blushing and disorder, tendered me from them a ridiculous paper which concerned most some particularities of China, and those Oriental parts."[60] The Society's apparent lack of knowledge of geography was the subject of Stubbe's jousting in *The Lord Bacons relation to the sweating-sickness examined,* in which he argued that the Royal Society's project of natural history was "not necessary to the World, except one have an occasion to send to the East-Indies to know what grows in America, or to Southwales for an account of Nova Zembla, or the Countries subject to the North and South Pole."[61] More biting was Stubbe's question as to "who can with any patience read how this famous Society sent to the Governor of Batavia in the East-Indies to know what grown in Mexico in the West-Indies?"[62] Although Stubbe's personal animosity towards the Royal Society obviously colours his writing, his sentiments illuminate the fact that, lurking beneath the frequently triumphal tone of the *Transactions,* boastful of new discoveries, there were a number of problems plaguing the process of accumulating knowledge. These issues are set in relief by the second practice of the Royal Society's New World empire-building; the collection of rarities.

IV: The Culture of Collecting

After almost a decade of sending entreaties to Connecticut, in the autumn of 1669 Oldenburg finally received what he had been hoping for. Several wooden boxes arrived from New England containing a myriad of natural curiosities. The sender was John Winthrop, who described the objects in an accompanying letter.

> There is in a broad round box a strang kind of fish wch was taken by a fisherman [in] ....Massachuset Bay in New England...There is in an other box a fish wch is full of prickles wch they call a seahedghog; as also a small flying fish...these flying fishes are chased by the dolphin, & that causeth them to fly out of the water...in those seas betweene these parts and the West-Indies....In the same box are heads of a vegetable we call silke grass...pieces of the barke of a tree wch growes at Nova Scotia...the eares of Indian cone...There is also put aboard

> [the ship] loose …the head of a deare wch seemeth not an ordinary head. It was brought far out of the country by some Indians.[63]

The second way in which the Royal Society attempted to recover man's epistemic empire in America was a very tangible one. It was the accumulation and display of specimens. In their drive for an encyclopedic representation of the natural world, a repository or museum represented a microcosm of the ideal of an empire of knowledge. But as Winthrop's long-awaited bounty reveals, there were occasionally difficulties involved in the practice of collecting. An examination of the Society's Repository and the private collections of its Fellows offers us an important insight into the limited success of realizing their vision of reclaiming man's original empire.

The Royal Society's collection of curiosities existed from its earliest days, in fact, before the society was formally incorporated in November 1662. In that year, they established a Repository at Gresham College London which was attended by Robert Hooke, who became known by the title "Curator of Experiments." In July 1678, the botanist Nehemiah Grew was asked to catalogue the repository. A year later the result was printed, and titled, *Musaeum Regalis Societatis, or, A Catalogue and Description of the Natural and Artificial Rarities Belonging to the Royal Society and Preserved at Gresham Colledge.*[64] Grew's choice of the word "museum" to describe the Repository is significant. The *Oxford English Dictionary* reveals that the word only began to be used in English in the 1650s. By naming the Royal Society's Repository a museum, Grew helped sanction the term for scientific enterprise.

In his catalogue, Grew attempted to put into practice the Adamic ideal of knowledge in which plants had names which corresponded to their natures; these names were not mere signs. He tells us that, where plants and animals did not have names, he named them himself. In doing so, he tried to name them based upon "some-thing more observably declarative of their Form, or Nature. The doing of which, would much facilitate and Improve the Knowledge of them. For so, every Name were a short Definition."[65] In the last few paragraphs of his Preface, Grew places the cataloguing project in a Baconian context in which natural specimens contribute to the usefulness and improvement of knowledge. "The greatest Rarity, if once experienced to be of good use, will soon become common."[66]

The American colonies played a central role in the project of collecting specimens. Many of the Society's most reliable correspondents were stationed in the colonies, and furthermore, the colonies" position on trade routes facilitated the trans-Atlantic shipping of artifacts. Approximately two thirds of the objects in Hans Sloane's personal collection, for example, had an American provenance, and it was Sloane's colonial activity as physician to the Governor of Jamaica which enabled him to gather the eight hundred plants and menagerie of animals which became the basis for his collection.[67]

The importance of the American colonies is also evident in the Oldenburg's requests to his correspondents, many of whom were in the Americas. Writing to Richard Norwood in the Bermudas, Oldenburg requested him "to send in what Observables you might meet." Richard Stafford, who arrived in Bermuda in 1626, and was briefly the sheriff there, wrote to inform Oldenburg that he should receive from Captain Thomas Morlie "the commandr: of our Magazieene ship such thing as I could at present procure." John Winthrop wrote frequently about "the collection of some of the productions of the wildernesse" that he made, and the "rarities."[68]

In spite of its Baconian idealism, the reality of the Society's Repository was rather different, primarily because of the Society's complex relationship with commerce. On the one hand, some Fellows, notably Robert Boyle, saw commerce with American colonies as central to the recovery of the empire of knowledge. This is because Boyle envisaged a fruitful exchange of knowledge between traders and naturalists. "The Experimental Philosopher may not only Improve Trades, but Multiply them."[69] Boyle idealized the "transplanting of Arts and Manifactures" from England to the colonies, and vice versa. "Experimental Philosophy", he wrote, "may not only it self be advanced by an Inspection into Trades, but may advance them too; so the happy Influence it may have on them is none of the least wayes, by which the Naturalist may make it useful to promote the Empire of Man."[70]

When it came to collecting, however, the influence of commerce on the Society and its members' endeavours was more ambiguous. Often the collaborative and professional aspirations of the Society's Repository were the exception rather than the rule. Indeed, the fact that some of the Royal Society's Fellows owned their own private collections somewhat undermined Bacon's injunction to create public, collaborative knowledge which he trumpeted in the *Parasceve.*

Elias Ashmole's story is also a case in point.[71] The son of a Staffordshire saddler, Ashmole rose through the social hierarchy by trying to ingratiate himself with the Order of the Garter. He also married wisely. His second wife, Lady Manwaring, was considerably wealthy. She was also twenty years his senior. The fact that one of her sons tried to murder Ashmole may well indicate the fact that his motives were showing. His shady marriage aside, the chief way in which Ashmole tried to promote himself was through his collecting. His collection, which we know today in the form of the Ashmolean Museum, originally belonged to the Tradescants, and it contained many American rarities. Arthur MacGregor's research into the surviving collections at the Museum has yielded a list of extant artifacts that can be traced to the 1685 catalogue. Items from the Americas include: a paddle; ball-headed clubs; non-ball-headed clubs; a self-bow; a beaded wampum belt ("wampum" being a colonial abbreviation of the Algonquian word "wampumpeake" which refers to white and purple tubular shell beads.) There was also a shirt made of skin, the Mantle of Powhatan, the Amerindian King of Virginia which is decorated with beadwork; a skin pouch, a jaguar-tooth pendant, a string of Ocelot teeth (Ocelot is a native wildcat); and a hammock. In addition, a number of ethnological specimens included Wampum beads; a model canoe probably from Canada; a south American Indian Club; a Hammock from British Guiana made of cords and twisted grass; a hammock from South America made from white cotton thread.[72]

From about 1650 onwards, we know that Ashmole and his first wife ingratiated themselves with the Tradescants. He subsequently played a role in the publication of Tradescant's museum catalogue *Musaeum Tradescantium,* and he was later—at least so he claimed—given Tradescant's collection as a gift. Mrs Tradescant, perhaps unsurprisingly, disagreed about this.[73] The Tradescants had put the effort into building the collection, but it was Ashmole who created from Tradescant's Ark an institution of considerable status. A new building was constructed for the museum in Oxford, and the *Museum Ashmoleanum* was officially opened by James Duke of York in 1683. Entry to the museum was open to the public.

The case of Ashmole illustrates that the culture of collecting in the late seventeenth-century did not always make it the ideal site for the creation of an empire of knowledge. While the Royal Society's Repository was an attempt to recreate an epistemological empire in microcosm, the cultural context of this way of imagining America placed real limitations on how much of Adam's empire could be recovered.

Conclusion

It is one of the strange ironies of British imperial history that the project of bringing the New World under England's intellectual and scientific aegis was so haphazardly pursued.

The idea of America which I have explored in this chapter offers us, I think, one way of understanding this irony. Natural philosophers of the Royal Society of London conceived of America as part of man's original empire of knowledge. There was a tension, however, between their grand hopes for recovering man's pre-lapsarian epistemic empire, and the less impressive material practices through which they gleaned knowledge of the New World. The Society's collection and shipment of colonial artifacts, and their efforts to compile an encyclopaedic natural history—despite brave attempts at Baconian meticulousness—did not realize their foundational ideal.

Here the "fit of absence of mind" of which John Seeley spoke so many decades ago, might give us pause for thought. The Royal Society's imagination of America in explicitly imperial terms challenges the theory that the English were absent minded imperialists. The point, I would suggest, is not a lack of imperial consciousness. Rather, it is the complex disjunction between the theory and practice of the idea of empire in the seventeenth-century. Understanding the imagination of America as a recovered empire of knowledge, therefore, might help us reorient our thinking about the conceptual history of the British Empire.

---

[1] I would like to thank Mark Goldie for his helpful comments on an earlier draft of this chapter. J.H Elliott, *The Old World and the New, 1492-1650,* (Cambridge, 1992 [1970]), 14.

[2] Richard Drayton, *Nature's Government: Science, Imperial Britain and the "Improvement" of the World,* (New Dehli, 2000); Joyce Chaplin, *Subject Matter: Technology, the Body and Science on the Anglo-American Frontier 1500-1676* (Cambridge, Mass., 2001); Patricia Seed, *Ceremonies of Possession in Europe's Conquest of the New World, 1492-1640* (Cambridge, 1995); Richard Grove, *Ecology, Climate and Empire: the Indian legacy in Global Environmental History*, (Oxford, 1998); Robert Iliffe, "Foreign bodies, Part Two: Foreigners and the making of the early Royal Society," *Canadian Journal of History* 53 (1999), 23-50; J.H Rowe, "The Renaissance Foundations of Anthropology," *American Anthropologist,* 67 (1965), 1-20.

[3] Genesis 1:28 (King James version)

[4] Daniel 12:4

[5] Henry Oldenburg to John Winthrop, 13 October 1667. A. Rupert Hall and Marie Boas Hall, trans., *The Correspondence of Henry Oldenburg,* (Madison, Wisc., 1965) Vol. 3, 525.

[6] *Philosophical Transactions of the Royal Society,* Volume 16, Number 180, March and April, 1686, 37.

[7] For a good discussion of the challenge to ancient texts posed by America's discovery see Anthony Grafton, *New Worlds Ancient Texts: the Power of Tradition and the Shock of Discovery,* (Cambridge, Mass, 1995).

[8] Francis Bacon, *The New Organon,* Book I, LXXII, ed. Lisa Jardine and Michael Silverthorne, (Cambridge, 2000), 60.

[9] Richard Yeo, *Encyclopaedic Visions: Scientific Dictionaries and Enlightenment Culture,* (Cambridge, 2001), 2.

[10] On the Hermetic philosophy see Frances Yates, *Giordano Bruno and the Hermetic Tradition,* (London: Routledge, 2002 [1964]), and on the occult tradition see Frances Yates, *The Rosicrucian Enlightenment,* (London, 2002 [1972]).

[11] Ralph Cudworth, *The True Intellectual System of the Universe* (London, 1678).

[12] Sabine McCormack, "The Relationship between Graeco-Roman and Amerindian Paganism in Early Modern Europe," in, Karen Ordahl Kupperman, ed., *America in European Consciousness* (Chapel Hill, 1995), 98.

[13] Ibid., 96.

[14] Amy G. Gordon, "Confronting Cultures: The Effect of the Discoveries of Sixteenth-Century French Thought," *Terrae Incognitae,* 8 (1976), 45-57; Peter Burke, "America and the Re-writing of World History," in Kupperman, ed., *America,* 33-51.

[15] See J.H Rowe, "The Renaissance Foundations of Anthropology," *American Anthropologist,* 67 (1965), 1-20; Anthony Pagden, *The Fall of Natural Man: The American Indian and the Origins of Comparative Ethnology* (Cambridge, 1982); Rudolf Pfeiffer, *The History of Classical Scholarship from 1300 to 1850* (Oxford, 1976).

[16] On these theories see Roy Porter, *The Making of Geology: Earth Science in Britain, 1660-1815* (Cambridge, 1977).

[17] William B. Ahsworth Jr, "Emblematic Natural History of the Renaissance," in Nicholas Jardine and Emma Spary eds., *Cultures of Natural History,* (Cambridge, 1996), 20.

[18] Scott Atran, *The Cognitive Foundations of Natural History: Towards and Anthropology of Science,* (Cambridge, 1990).

[19] See for example Henry Lowood, "The New World and the European Catalog of Nature" in Kupperman, ed., *America,* 295 – 323.

[20] Michel Foucault, *The Order of Things: the Archaeology of the Human Sciences* (New York, 1970)

[21] Peter Harrison, *The Bible, Protestantism and the Rise of Natural Science* (Cambridge, 2001)

[22] Marjorie Swann, *Curiosities and Texts: the Culture of Collecting in Early Modern England* (Philadelphia, 2001), 18.

[23] Cited in Ibid., 23.

[24] *Calendar of State Papers Colonial Series,* 1, no.11.

[25] Arthur MacGregor, "John Tradescant."

[26] John Prest, *The Garden of Eden: the Botanic Garden and the Recreation of Paradise* (New Haven, 1981), 27.

[27] Ibid., 31.

[28] Ibid., 33.

[29] Drayton, *Nature's Government,* 50. See esp. 50-82.

[30] Seed, *Ceremonies of Possession.*

[31] John Locke, *Two Treatises of Government* ed., Peter Laslett, (Cambridge, 1963) 2, V, §27. On John Locke and colonialism, see David Armitage, "John Locke, Carolina and the Two Treatises of Government" *Political Theory,* 32.5 (October 2004), 602-27; Duncan Ivison, "Locke, Liberalism and Empire" in *The Philosophy of John Locke: New Perspectives,* ed. Peter Anstey, (London, 2003), 86-105; Barbara Arneil, "Trade, plantations and property: John Locke and the economic defence of colonialism" *Journal of the History of Ideas,* 55 (1994), 591-609; Barbara Arneil *John Locke and America: the defence of English colonialism* (Oxford, 1996). Herman Lebovic, "The Uses of America in Locke's Second Treatise of Government" *Journal of the History of Ideas,* 47.4 (October 1986), 567-581; James Tully, "Rediscovering America: The *Two Treatises* and Aboriginal Rights" in *An Approach to Political Philosophy: Locke in Contexts,* (Cambridge, 1993), 137-76.

[32] *Philosophical Transactions,* Volume 9, 1674, epistle dedicatory, unpag.

[33] Robert Boyle, "A Disquisition about the Final Causes of Natural Things: Wherein it is Inquir'd Whether And (if at all) With what Cautions a Naturalist should admit them." *The Works of Robert Boyle,* ed. Michael Hunter and Edward B. Davis, (London, 2000) v.11, 108. Original emphasis.
[34] "Of the Usefulness of Mathematics to Natural Philosophy Or, That the Empire of Man may be promoted by the Naturalist's skill in Mathematics, (as well Pure as Mixt.)", Usefulness II, Volume VI, 441.
[35] "Usefulness of Natural Philosophy", II, 2, Section 7, Volume VI, 435.
[36] Joseph Glanvill, "Essay on The Usefulness of Real Philosophy to Religion" *Essays,* 41.
[37] "Christian Virtuoso II", Volume XII, Aphorism V, 444-5.
[38] *Philosophical Transactions,* Volume 7, 1672, general dedication for the volume, unpag.
[39] Henry Oldenburg to Rene Francois Sluse, 23 October, 1667, *Oldenburg Correspondence,* Volume 3, 537.
[40] Ibid., 535-6.
[41] *Philosophical Transactions,* Volume 17, Number 203, 1693, 978.
[42] *Philosophical Transactions,* Volume 19, 225.
[43] *Philosophical Transactions,* Volume 5, Number 57, 25 March, 1670, 1151.
[44] John Locke to Henry Oldenburg, 20 May, 1675, *Oldenburg Correspondence,* Volume 11, 2667
[45] Martin Lister to Henry Oldenburg, 27 June, 1673, *Oldenburg Correspondence,* Volume 11, 373.
[46] *Philosophical Transactions,* Volume 7, Number 83, May 20, 1672, 4078.
[47] Hans Sloane, *Voyage to the Islands of Madera, Barbados, Nieves, S Christophers and Jamaica, with the natural history of the last of those islands; to which is prefixed an introduction, wherein is an account of the inhabitants, air, waters, diseases, trade &c.* (London, 1707-25) Preface.
[48] Ibid.
[49] Henry Oldenburg to John Winthrop, 11 April, 1671, *Oldenburg Correspondence,* Volume 7, 569.
[50] Francis Bacon, *Parasceve, or Preparative towards a Natural and Experimental History, affixed to the Novum Organum.* James Spedding, R.L Ellis, and D. D Heath, (eds) *The Works of Francis Bacon,* (London, 1860)
[51] *Philosophical Transactions of the Royal Society,* Volume 1, Number 8, January 1665-6, 141.
[52] *Philosophical Transactions,* Volume 1, Number 11, April 2, 1666, 188.
[53] Ibid.,187.
[54] *Philosophical Transactions,* Volume 11, Number 23, March 11, 1666/7, 420-1.
[55] Ibid., 422.
[56] *Philosophical Transactions,* Volume 3, Number 33, March 16, 1667/8, 634.
[57] Ibid., 634.
[58] Ibid., 634.
[59] Ibid., 634-9.
[60] Ibid., unpag.

[61] Henry Stubbe, *The Lord Bacons relation to the sweating-sickness examined, in a reply to George Thomson, pretender to physick and chymistry together with a defence of phlebotomy in general, and also particularly in the plague, small-pox, scurvey, and pleurisie, in opposition to the same author, and the author of Medela medicinae, Doctor Whitaker, and Doctor Sydenham: also, a relation concerning the strange symptoms happening upon the bite of an adder, and, a reply by way of preface to the calumnies of Eccebolius Glanvile* (London, 1671), LI 2.
[62] Henry Stubbe, *Legends, no histories, or A specimen of some animadversions upon The history of the Royal Society,* (London, 1670) unpag.
[63] John Winthrop to Henry Oldenburg, 4 October, 1669, *Oldenburg Correspondence,* Volume 6, pp.256-7.
[64] Nehemiah Grew, *Musaeum Regalis Societatis, or, A Catalogue and Description of the Natural and*
Artificial Rarities Belonging to the Royal Society and Preserved at Gresham Colledge (London, 1686)
[65] Ibid., Preface, unpag.
[66] Ibid., Preface, A.
[67] Christian F Feest, "The Collecting of American Indian Artifacts in Europe 1493-1750" in Karen Kupperman, *America in European Consciousness,* 336.
[68] Henry Oldenburg to Richard Norwood, 24 October, 1666, *Oldenburg Correspondence,* Volume 3, 276. Richard Stafford to Henry Oldenburg, 16 July, 1668, *Oldenburg Correspondence,* Volume 4, 552. John Winthrop to Henry Oldenburg, 4 October, 1669, *Oldenburg Correspondence,* Volume 6, 253.
[69] "Usefulness II" Volume VI, 424.
[70] Ibid., 483.
[71] On Ashmole see Vittoria Feola, "Elias Ashmole and the uses of Antiquity" (Ph D diss., Cambridge University, 2005).
[72] Arthur MacGregor, *Tradescant's Rarities: Essays on the Foundation of the Ashmolean Museum, 1683, with a Catalogue of the Surviving Early Collections,* (Oxford, 1983),108-139
[73] On Ashmole, see A.G. MacGregor and A. J Turner, "The Ashmolean Museum" in, *The Eighteenth Century,* L.S Sutherland and L.G. Mitchell, (eds.,) vol 5 of the *History of the University of Oxford,* (Oxford, 1986); Conrad H. Josten, *Elias Ashmole: His Autobiographical and Historical Notes, his Correspondence, and Other contemporary Sources Relating to his Life and Work* (Oxford, 1966)

Chapter Three

British Imaginings of the Eighteenth-Century Southeast

Ian Chambers

> He would have guarded the door himself and put to death every one that endeavored to make their Escape.[1]

Since the 1580s when the English first landed in North America the southeastern Native population has been seen and imagined in a variety of ways. By 1607, and the founding of Jamestown, English perceptions had developed into two interlocking views. The first saw southeastern Natives as helpful and generous while the second viewed them as devious and dishonest. However, for both groups "lurking beneath the surface" was the "anticipation of violence," a belief that would be confirmed by the 1622 Powhatan confederacy attack on Jamestown.[2]

From the founding of Jamestown, and more explicitly the development of tobacco as a staple crop in the seventeenth century, the British increasingly began to envisage the western coast of the Atlantic Ocean as a location for permanent settlement. By the early eighteenth century, the British population had been exposed to over a century of tales from the new colonies, had pored over maps of the region, and on several occasions had the opportunity to encounter southeastern Native Americans when they visited Britain.[3] This increasing knowledge of southeastern space and its people combined with the "growing quantifying spirit of the eighteenth century" led to a belief that the Native people of the southeast, now well known, and the space they inhabited could, and importantly should, be subjected and incorporated within the British Empire.[4]

This chapter examines this belief through the activities of one individual's experience. The story begins not in the hills and mountains of the southeast but rather in the city of Edinburgh, Scotland, where on December 18, 1691

Alexander Cumming was born.[5] Twenty-eight years later in 1729, Cumming "was induced, by a dream of Lady Cumming, to undertake a voyage to America, for the purpose of visiting the Cherokee nations."[6] From these peculiar beginnings came one of the strangest moments of British-Cherokee interaction that ended with seven Cherokees in the court of King George. This is the view of a British visitor who only briefly visited Cherokee space, indeed long-term traders were both shocked, and fearful of the response, to his actions. His interaction with, and view of, the Cherokee was pursued with what one historian refers to as "tourist enterprise," and took place over a four-week period in 1730. Cumming is an ideal subject to explore the British imagining of the Cherokee.[7]

While the full manuscript of Cumming's Journal is now unfortunately lost, it was published in two sections in *The Daily Journal*, a London newspaper, upon his return to England. [8] The first section, published in September, 1730, was an overview, detailing in broad strokes the daring deeds of Cumming leading up to his arrival in London with seven Cherokees, claiming to be the appointed leader of the Cherokee nation. The second report, published the following month, gives a more detailed day by day description of his travels.[9] Walking with Cumming as he took his journey into Cherokee space allows us to uncover the moments where the British imagination and understandings of southeastern Native Americans were used to inform understanding of the unfolding events. This chapter will also show how the Cherokee reacted to, and made sense of Cumming's arrival.

In a period when the contest for control of what Herbert Eugene Bolton has termed the "Debatable land" of the southeast any statement of spatial control was viewed not only as a geographical but also as a political success.[10] The publication of Cumming's Journal, and the widely publicized presence of the Cherokee who accompanied him, reassured the British public of their country's position of strength in the region and also allowed the British government to solidify its relationship with the Cherokee. The combination of public and political recognition assured that in the eighteenth century the trip was seen as a moment of great success for British colonial ambitions.

Over time Cumming's position has been interpreted differently. During the nineteenth century historians ascribed Cumming the role of official ambassador from the British crown, something that Cumming himself had denied in his journal.[11] More recently, he has been viewed as nothing more that an oddity, "a one-man firework display wildly emitting sparks and colored lights and blazing rockets" whose arrival coincided with the earliest steps towards contemporary Cherokee state formation.[12] I propose to re-introduce Cumming as an important part of early eighteenth century Cherokee-British interaction. To show a moment when the British imagining of the southeastern Native American population prompted an attempt at political and spatial domination. An attempt concurrently viewed by the Cherokee through their own interpretation of the other, in this case the British. A moment when differing interpretations of colonial interaction are brought into sharp relief.

Arriving in Charles Town in December 1729 and using his title as collateral, Cumming convinced the colonists that he possessed great wealth and proposed to "settle in Carolina, and do wonderful Things for the Good of the Country."[13] To solidify his position, and signal his intent to assist the people of the colony, he began to write promissory notes which "by his punctual Payment of them upon Sight, in a little Time they acquired a Credit and Currency equal with Money." With this secure footing, Cumming began to expand his financial dealings within the colony, buying "several large Plantations." In addition he opened a loan office, issuing more notes bearing a ten-percent interest payment. Using the money gained from these several ventures, he was able to buy an "abundance of Gold and Silver and a great Deal of Country Produce which he shipt away continually." With his financial and social position secured, Cumming was able to draw on the help of several of the leading traders and merchants to furnish his journey among the Cherokee.[14]

On the afternoon of March 13, 1730, at "about five," Cumming left the home of Mr. James Kinlock at New Gilmerton, South Carolina, in the company of Indian trader George Chicken and Surveyor George Hunter, and headed into Cherokee country. Three days later, Cumming and his companions, who now included William Cooper, "a bold Man, and skilled in the Cherrokee Language", traveled to an underground cave near the home of a Mr. Coxe. Here is an early indication of Cumming's desire to leave his mark in South Carolina, in this case literally. [15] Cumming informs us that "Mr. Chicken, and Mr. Coxe, made several marks to show that they had been there." Not content to enter just his own name, Cumming branded the cave, and symbolically the land, for Britain engraving "King George II of Great Britain, wrote by S.A.C."[16] By leaving this mark, Cumming was attempting to ensure that the space in which he stood became known not only physically but was also conceived of as part of British jurisdictional and intellectual space. This desire to identify, fix and record is seen throughout the journey. Whenever Cumming met an "Indian" he would "take his name down in his book saying that he had made a friend of him." By this act Cumming has added each individual to the knowledge and control of the British.

Over the next few days Cumming continued his journey along the Cherokee Path, the main trading trail into Cherokee Country. [17] Arriving at the Cherokee town of Keowee on March 23, 1730, the true purpose of his journey was revealed to both the Cherokee and his traveling companions. Over dinner in the house of one of the town's traders, Cumming was informed that there recently had been "Messengers from the lower Creeks, with the Cherokees, desiring them to come over to the French Interests."[18] Cumming suggests in his journal that this warning reinforced the danger that he was under and the bravery he showed in his journey. However this warning provided the catalyst that forced Cumming's next action. As a self-conceived ambassador, acting, in his own mind, on behalf of Britain to visit, define, mark, and thereby control the Cherokee nation, Cumming was forced to act by the threat of French incursion into the region. His response to the news is demonstrated in the events that unfurled in the Keowee Town House that evening.[19]

The Town House was full, with a reported 300 hundred Cherokees in attendance as well as nine British traders. Cumming entered and waited patiently while the Town's leaders spoke before making his own speech.[20] Cumming began by informing the Cherokee who he was and that he had come as an individual citizen, not as an official ambassador on behalf of King George. He then proposed a toast to the King during which he hoped, if not demanded, that all others would pledge their loyalty and allegiance to King George. After encouraging the shocked traders to fall to their knees, he turned to the Cherokee and, revealing the four firearms and the cutlass he had under his cloak, demanded that the "head Warriors… acknowledge his Majesty King George's Sovereignty over them on their knee." Joseph Cooper, a trader who was the evening's interpreter also, declared:

> If he had known before hand what Sir Alexander would have order'd him to have said, he would not have ventured in the Town-House to have been Interpreter, nor would the Indian Traders have ventured to have been Spectators, believing none of them could have gone out of the Town-House without being murdered, considering how jealous that People had always been of their Liberties.[21]

Cumming and his contemporaries leave us no description of the Cherokee reaction to this alarming episode, although as all the British lived through the event we must assume that some form of accommodation was met. For Cumming, situated within the British spatial and political understanding, the meeting represented the conquest, or at least the first step in that process, of the Cherokee nation for the British crown. In Cumming's mind, the Town House was no longer a location for the meeting of Cherokee people to discuss and debate the affairs of their Town and through clan affiliation the broader Cherokee nation; the space was now the site where British political authority and control had been established. The Town House and the space it occupied now personified Cumming's, and through him Britain's, control of this Cherokee village. Cumming viewed the population of the village as British subjects, informing them that if they failed to submit to the king "they would become no people," for any population beyond British space had no fixidity and therefore no existence.

Cumming's intention to dominate the Cherokee is seen by his response to trader Ludovick Grant. When asked what he had planned to do if the Cherokees present had refused to submit:

> he [Cumming] answered with a Wild look, that . . . if any of the Indians had refused the King's health to have taken a brand out of the fuire that Burns in the middle of the room and to have set fire to the house. That he would have guarded the door himself and put to death every one that endeavored to make their Escape that they might have all been consumed to ashes.[22]

This statement, which worried the traders who heard it, indicates that for Cumming there were only two positions available to the Cherokee, to accept their position as subservient members in British space or death.

To affirm his position and his claim of dominance over the Cherokee, Cumming demanded a second meeting eleven days later, on April 3, 1730, at the Cherokee Town of Nequassee. Cumming informed the members of the Keowee community that he expected that "one of their head Men should bring full Power from the lower Settlements, another full Powers from the upper Settlements, and the third full Powers from the middle settlements."[23] In demanding individuals to represent the physical regions of the whole nation, Cumming planned to repeat his conquest of the single village on a national scale.

For those Cherokee in the Keowee Town House on March 23, 1730, operating in a different political understanding, an alternative interpretation of the events would be realized. James Adair, a contemporary of Cumming, provides us with a clue as to the Cherokee understanding. Writing of Cherokee government Adair notes that:

> They can only persuade or dissuade the people, by their force of good-nature and clear reasoning, or colouring things, so as to suit their prevailing passions. It is reputed merit alone, that gives them any titles of distinction above the meanest of the people.[24]

What Adair saw as a lack of authority within the administration of the Cherokee was actually a recognition that each person's voice and opinion could, and would, be heard in debate. This belief in an egalitarian approach to politics did not mean, however, that every point contained equal value. As the quotation suggests certain people could gain a greater degree of influence through their action or "merit." This opens up the possibility for someone with a proven history of success, in warfare or negotiation, to rise into a position of heightened influence, what Adair referred to as a "title of distinction."

Adair's quotation also suggests that strength of argument and oratorical skills played a large part in any debate. It is therefore possible that an individual could, through force of personality and argument, influence the policy of the Town. Accepting that all people could be heard and that a powerful debater could gain influence, is it possible that this was the route that allowed Cumming to, apparently, dominate the community of Keowee. Cumming's behavior in Charles Town prior to his journey into Cherokee space indicates that Cumming had self-confidence and skills of persuasion, and his willingness to make such a dramatic display of force in the Keowee Town House suggests that he possessed tremendous bravery and assumed great authority. Both of which may have allowed him to assert power within the space of the Town House.

If we view the episode at Keowee as an isolated and individual event, the apparent submission of the Cherokee appears puzzling, even allowing for Cumming's oratory skills and assertive behavior. However, if the events of the evening of March 23, 1730 are placed into a broader view, based within a Cherokee understanding, a different image begins to appear. Although Europeans reported, and Cumming believed, that by his actions he had ensured control and domination over the Cherokees, in actuality Cumming had been absorbed or adopted into Cherokee space.

The first step in understanding this alternate view of Cumming's involvement with the Cherokee focuses on the role of negotiation within Cherokee political life. Cherokee individuals involved in negotiations had authority, in a loose sense, to represent the position of the larger group but no authority to enter into additional negotiations or agreements without further consultation with the larger group. Cumming's actions, therefore, must be reinterpreted in light of this. Whereas Cumming saw his request at Keowee to summon other Cherokee leaders, as an instruction to facilitate a broader control, the Cherokee recognized it as an opportunity to adjourn the negotiations and return to the larger group to discuss the next step. For the Cherokee, the single Town of Keowee did not have the authority to commit the whole nation to a position of allegiance. The apparent Cherokee acquiescence in the Keowee Town House on March 23, 1730, was in actuality a necessary step in their political structure, that is, they deferred the negotiations until a larger Cherokee group was able to convene and discuss Cumming's proposal.[25]

Working in tandem with the need for larger group discussion was a second aspect of Cherokee political life, the need for harmony. The Cherokee spatial and political system had at its core the recognition that discrete places needed to be combined through ceremonial means for life to continue. Within political debate, this aspect of the spatial persona played out through the pursuit of harmony. When harmony was threatened, the deferral of decision making was a common method of maintaining harmony. When involved in negotiations the

Cherokee deflected requests for immediate resolutions instead choosing to withdraw, discuss, and return.

An example of the manner in which pursuit of harmony through deflection operated can be found by turning to another of Cummings's contemporaries. On Tuesday January 14, 1725 at the Town of Tugoloo, Colonel George Chicken met with representatives from several Cherokee Towns. During the meeting Chicken put forward seven points for discussion, which ranged from details of recent activity among the Creek to the Cherokee relationships with the English. The fifth point raised by Chicken, detailed below, marked the beginning of the Cherokee process of decision deferral. Chicken raised the subject of a standing Cherokee army. He informed the Cherokee that:

> if you would but Consider Yourselves how Numerous you are and how little you would Miss the drawing out of each Town in the Nation a small Number of Men, you would not talk of defending your Towns but would Raise an Army of Men and Defend your Enemies before they come Nigh your Towns.[26]

As the above quotation shows Chicken's British origins led him to assume that the logic of a single unified national project, such as an army, would make sense and be acceptable to the Cherokee. For the Cherokee, however, the need for separate discrete spaces, such as Towns, each responsible for their own defence, was the logical answer to the challenge of external enemy attack. In order to preserve harmony within negotiations where British and Cherokee views were clearly divergent, and not going to be easily resolved, the assembled Cherokee leaders attempted to defer a decision assuming that Chicken would recognize and accept this tactic.

Initially the Cherokee promised to "all go and consult together" promising to return later with an answer. Upon further discussion the Cherokee informed Chicken that they would take action upon his suggestion to send out scouts to look for the enemy, but that with regard to the formation of a national Cherokee army, they "had concluded to send to the other Towns in order to meet them." Chicken's response to this deferment was once more to push the concept of a unified army. He informed the Cherokee that "unless they had a body of men to go out against the enemy when they were discovered that their Scouts would be of little Service to them." The Cherokee continued to defer a decision, while Chicken again pushed for the creation of an army.

The Cherokee then attempted a different approach to defuse and deflect the problem. They informed Chicken that they accepted that "if the enemy comes on them before they can gett a body it would not be the Englishes fault because they have given them Notice of it." Even with this assurance that blame would not be apportioned to the English, an important concession in a clan based society like the Cherokee where culpability was a key factor in decisions of retaliation. Despite this assurance Chicken still refused to back off and once more pushed for a commitment to a standing army. In a final attempt to deflect the decision, driven, we must assume, by exasperation at Chicken's continued failure to recognize the correct form of negotiation, the Head Warrior of Toxsoah agreed to go out and "gett what men he could to goe along with him." This final act allowed for an acceptable solution to both parties. Chicken assumed that this was the beginning of a national army with the Head Warrior of Toxsoah as the commander, while the Cherokee saw it as the promise of one man, not of the whole group. Thus after several hours of negotiation, harmony was restored.[27] The use of decision deferment was one intellectual tool available to the Cherokee during their interaction with Cumming. A second involved the Cherokee relationship to personal power.

Individual power for the Cherokee was not a fixed attribute possessed inherently. Rather, it was a potential that the correct behavior could reveal and incorrect actions could remove. Power was therefore flexible and no one was able to assess instantly how much power an individual might have. Thus "overt deference and respect afforded the safest course to follow in interpersonal relations. Such behavior minimized the chance of giving offense and, perhaps, suffering hostile repercussions."[28] Cumming's actions, bringing weapons into the Town House and demanding subservience, put him outside the normal rules of behavior for Cherokee negotiations. It made eminent sense for the Cherokees to defer a decision and palliate someone behaving so aggressively. Yet, these same actions suggested that Cumming possibly possessed personal power or *ulanigvgv*.[29]

An event that occurred between Cumming's stance in the Town House of Keowee and his later "coronation" at Nequassee suggested, from the Cherokee view, that Cumming may have possessed great individual power. On March 27, as Cumming entered the Town of Tassetchee "there happen'd to be most terrible Thunder, Lightning, and Rain."[30] One European who had lived with the Cherokee for ten years in the early eighteenth century reported that Cherokee perceived thunder and lightening as bringing messages from spiritual beings. Individuals who assumed themselves to be superior to others and gave "themselves over to al sort of crulletie and abominations" would be struck down by lightening as punishment for their inappropriate behavior. However, if lightening did not strike the offender but fell "hard by Them," then "that is a message" for others to "ammend thire lives and To obay thire seperriors."[31] Cumming's decision to take guns into the Town house and demand the acquiescence of the Cherokees present would appear to locate him as one claiming undue prestige and superiority. Yet a Cherokee reading of the meteorological events of the evening of March 27 when Cumming remained unscathed by the thunder and lightening suggested that Cumming possessed strong spiritual power.

Cherokee caution regarding power combined with their preference to defer decisions when unanimity could not be reached suggest another way to understand the events involving Cumming between March 23 and April 3, 1730. Seen from a Cherokee perspective, Cumming neither controlled nor dominated the proceedings. Instead, Cherokees constructed a meaning and a place for Cumming that allowed him to be respected but not accepted. They recognized the potential Cumming held but did not ascribe him authority. Rather, the Cherokees found a way within their political understanding to defer any decision making to a later date.

After leaving Tassetchee, Cumming headed to Great Tellico where he unwittingly describes an alternative set of political actions being played out. The actions taken by "Moytoy, the Head Warrior" shift Cumming's journey from a daring and aggressive act of British politics to place the Cherokee within British space, to a move by the leader of one Cherokee Town to politically adopt Cumming, thereby creating a link between the Cherokee and the British. In the process Cumming moves from being a King to being a pawn of Cherokee politics.

Arriving at Great Tellico on March 29, 1730, Cumming comments that he saw "a great many Enemies Scalps, brought in and put upon Poles at the Warriors doors."[32] This is his first mention of the martial prowess of the Cherokee. Two interpretations can be made of this display. The first would be to suggest that Cumming, influenced by the scalps, saw the Town and its leader as a dominant force in Cherokee society. That he would later position Moytoy as "Emperor of the Cherokee" reinforces this view. The second interpretation is the possibility that the scalps had been intentionally displayed by the Town not to assert Moytoy's primacy in Cherokee society but to suggest his strength as a warrior. As a warrior Moytoy was situated in the Cherokee space that dealt with diplomatic relations with outsiders, thereby offering himself as a suitable person to be the focus of interaction between Cumming and the broader Cherokee polity. Cumming saw an opportunity to assert British domination and control over the Cherokee, whereas the Head Warrior, Moytoy, saw an opportunity to create a link between the British and the Cherokee through Cumming. It appears that both parties viewed the meeting through the lens of their own political understanding.

During an initial meeting with Moytoy in the Great Tellico Town House, Cumming is informed that there had been discussion the previous year "among the several Towns" of the Cherokee to make "Moytoy Emperor over the whole, but that now it must be whatever Sir Alexander pleased."[33] With this statement Cumming drops his first hint that he personally will assume a position of dominance among the Cherokee, for if the man whom the Cherokee have already considered for leadership is prepared to submit to Cumming, then what apart from total dominance are the British to expect?[34]

The events that unfolded are, however, given a different light when viewed through the Cherokee eyes. Any power that Moytoy claimed was not to position himself as Principal Chief among the Cherokee. Rather he sought to place himself as the man who could control the potential power that Cumming possessed and by that control form a link between the Cherokee and British.[35] The following day, March 30, while still in Great Tellico, Cumming was involved in another Cherokee ceremony, writing that he was "Particularly distinguished in the Town-house by Moytoy, where the Indians sung songs, danced, and stroked his Head and Body over with Eagles tails." This event was not included in Cumming's first published overview of his trip where he included instead the later meeting at Nequassee on April 3, thereby highlighting the perceived coronation. However, for the Cherokee the March 30 meeting had great importance, for it provided the first step in linking Cumming and the Cherokee. By ensuring that the ceremony took place in the Town House of Great Tellico, and by assuming a leading role in the ceremony, Moytoy placed himself in the center of this process and positioned himself to act as spatial ambassador of the Cherokee.

The next major incident, and for Cumming the culmination of his ascendancy to leadership of the Cherokee, was the bestowing of the crown of Tannassy during his alleged coronation. This occurred on April 3, 1730, at the Town of Nequassee, where the "Kings, Princes, Warriors, Conjurers, and Beloved Men were all met."[36] Cumming described the events in the Nequassee Town House in the following terms:

> Here with great Solemity Sir Alexander was placed in a Chair, by Moytoy's Orders, Moytoy and the Conjourers standing about him, while the warriors stroked him with 13 Eagles tails and their Singers sung from Morning to Night, and, as their Custom is on Solemn Occasions, they fasted the whole Day.[37]

Cumming informs us that he gave a speech during the coronation proceedings in which he detailed the power of the King and demanded the Cherokee's submission. Additionally he "ordered that the Head Warriors should answer for their Conduct of the People to Moytoy, whom he appointed their Head, by the unanimous Consent of the whole People."[38] For Cumming and his readers the image is clear. The Cherokee at this moment became subjects of Britain and if not totally controlled were at least no longer politically outside the British political space. In order to give physical presence to the events, Cumming collected not only the signed affidavits of the British traders present but in addition "as Evidence of the Truth of what had happened" brought to London seven Cherokee.[39]

Cumming's actions on, and reaction and comments to, the events of April 3 indicate the manner in which a British spatial understanding, based upon fixidity and control, operates within a political act.[40] Cumming reached out from British space, looking for an opportunity to act *upon* the Cherokee, not to interact *with* them. At the first opportunity Cumming used force in an attempt to dominate and control the Cherokee, and as he continued his journey, he attempted forms of control, meeting Cherokees and recording names, fixing them in the record of his journey. Buoyed by early success he pushed his reach further, demanding a meeting with the leaders of the whole nation. At this later meeting at Nequassee, Cumming considers himself to have been crowned king of the Cherokee, thereby both securing his position and more importantly fixing the Cherokee nation within the British imagination. One of his first acts is to dictate the structure of society and outline the political relationship between the British and the Cherokee, ordering "the head Warriors should answer for the Conduct of their People to *Moytoy*, whom he appointed their head . . . and he to answer to Sir *Alexander*."[41]

Historian Verner W. Crane correctly argues that the Cherokee "had no real notion of acknowledging English 'sovereignty,'" much less of parting with their lands to either Cumming himself or to the "Great Man on the other side of the water." Nevertheless, it is clear that something of a political and ceremonial nature did occur on this date. The Cherokee did meet in national council and Cumming was involved in some form of ceremony in front of the assembled Cherokee. If the events do not portray Cherokee submission to British authority then what do they represent?

Firstly, let us examine the importance and significance of the national council at which Cumming claims to have obtained sovereignty. The dramatic ceremony that Cumming viewed as a coronation may not be as dramatic or as singularly important as Cumming would have us believe. In 1725, five years before Cumming's journey, Colonel George Chicken traveled throughout the Cherokee nation.[42] Chicken was not there to assert his control over the Cherokee but rather to bring a message or "talk" from the South Carolina authorities. Like Cumming, Chicken spent only a short time in Cherokee space traveling from Town to Town before speaking to a gathering of all Towns. Chicken's actions contradict Cumming's claim that the meeting surrounding his coronation was such a gathering as "never was seen at any one time in that Country." In fact it was not unusual for the Cherokee to gather in large numbers to hear the "talks" of British visitors. Chicken's 1725 journey provides another inconsistency in Cumming's claims. Chicken describes his own arrival in Keewhohee (the Town Cumming referred to as Keowee), in the following manner:

> At my Arrival here King Crow and the head men were out of the Town at their Plantations and a Messanger being sent to inform them of my Arrival, they Imediatly[sic] Repaired to the Town and soon after, they after their Ceremonial way placed me in a Great Chair in the most Publick Place in the Town and set down by me themselv's faning me with Eagles Feathers.[43]

The similarity between this ceremony and that which Cumming experienced in Nequassee is obvious. Therefore, we can suggest that the ceremony Cumming experienced was not as he thought, a coronation, but rather an often-used method to incorporate an external individual. It represented a political adoption which allowed the Cherokees to form a link to an external discrete space. Thus allowing the voice of that outside space to be heard Additionally, Cumming's positioning of Moytoy as "Emperor" in order to verify British control over the Cherokee is also open to a Cherokee interpretation. Moytoy's efforts can also be seen as his attempt to place himself as the gatekeeper to contact with the British. Moytoy did not declare himself sovereign over the Cherokee, although the British would increasingly view him as if he had, but that he offered himself as the person who would sponsor Cumming's interaction and linkage with the Cherokee nation.

The final point to be made with regard to the ceremony of April 3, 1730 concerns the artifact that gave Cumming's claim of coronation its greatest legitimacy, the crown of Tannassy.[44] Cumming, in the published details of his journey, mentions the crown and ceremony simultaneously tying together the ceremony and the crown. However, a closer reading of the events reveals that Cumming did not receive the crown on the day of the ceremony, as his journal implied, but on the following day, suggesting a separation between event and artifact. There are other inconsistencies in the tale of the crown and how it came into Cumming's possession. The quotation below, from Cumming's initial overview report of the journey, details Moytoy's and the Cherokee's submission to Cumming and positions the crown as a symbol of that capitulation.

> *April* 4. The Crown was brought from great *Tannassie*, which, with five Eagles tails and four Scalps of their Enemies, Moytoy presented to Sir *Alexander*, impowering him to lay the same at His Majesty's Feet.[45]

For the reading audience, this is a clear assertion that the Cherokee are within the British spatial and political knowledge as a people who are known and importantly controlled by the British. The next quotation, taken from Cumming's later day-by-day account, offers a subtle but important shift in the role of the crown.

> *April* 4. The Solemnity continued, Sir *Alexander* made some Presents, received their Crown, Eagles tails, and Scalps of their Enemies, to be laid at his Majesty King *George's* Feet.[46]

The crown has shifted position. It is no longer offered to Cumming as a symbol of obsequiousness but rather as a gift given after and in return for, "Presents" given to the Cherokee by Cumming. This subtle movement in proceeding has turned the object from a symbol of authority to a symbol of reciprocal contact.

Examining the reports of other Europeans present at the presumed coronation ceremony further complicates the role of the crown. In his journal, Cumming suggests that as early as March 30 the possibility of his coronation was raised, writing that Moytoy and Jacob "determined to present him [Cumming] with the crown of Tannassy."[47] This early introduction of the supposed coronation reinforces Cumming's later claim of authority. A different explanation comes from trader Ludovick Grant, who explains the crown's introduction in the following terms:

> From Telliguo we rode over to Tannassee and afterwards returned by Neguasse Where several Traders met us and a good many Indians. Sir Alexander had been informed of all the ceremonies that are used in making a head beloved man, of which there are a great many in the nation. They are called Ouka, and we translate that word king, so we call the Cap, he wears upon that occasion his Crown . . . Sir Alexander was very desirous to see one of them, and there being none at that Town One was sent for to some other Town. He Expressed Great Satisfaction at Seeing of it, and he told the Indians that he would carry it to England and give it to the great King George.[48]

Grant, who was with Cumming in both Great Tellico and Nequassee, indicates that, rather than being discussed prior to the ceremony of April 3, the crown was introduced at a later date. Additionally, it appears that the Cherokees had no intention of presenting the artifact to Cumming as a crown or any symbol of authority, but rather that it was an ethnographic artifact displayed only upon Cumming's insistence.

Finally we must consider the role that physical geography played in the events that occurred during Cumming's journey, exploring the role that different spaces played in the Cherokees interaction with Cumming. The claimed coronation takes place in Nequassee and Cumming asserts that it is Moytoy, of Great Tellico, who performs the act. The crown itself comes from a third discrete Town location, Tannassee. This diversity of geographical location involvement in the interaction between the Cherokee and Cumming is expanded following Cumming"s supposed coronation. On April 6, 1730, during his return to Charles Town, Cumming informs us that they "proceeded to *Ookunny*, where

Sir *Alexander* found a House built for him."[49] Cumming will later claim that the house came with "certain Territories there unto belonging as an acknowledgement" of his position "as their Governor & Lawgiver," stating that additional space was assigned to him "in every Town through which Sir Alexander had passed in his journey throughout the mountains."[50] This act, the linkage between landholding and political power, would appear logical in Cumming's spatial understanding, as it matched the associations seen within the system of hierarchical and political control that operated in the Britain at the time.

The act of constructing a house for Cumming adds a further aspect to the Cherokee understanding of Cumming's visit. By constructing a house for Cumming, in a Cherokee Town, Ookunny, and in a Cherokee style, the Cherokee have positioned and included Cumming in their physical and intellectual space. They have created a way for Cumming to have a stake in one of the two aspects of Cherokee national membership: Town affiliation.

The positioning of Cumming's residence in Ookunny, which appears to be a village with ties to the Township of Keowee, achieves two further objectives.[51] Firstly, it places Cumming at the edge of Cherokee space in the region most closely related to Charles Town, thereby increasing the ease of communication between British and Cherokee space. Secondly, it offers the intriguing possibility that Cumming may have acquired some form of national Cherokee membership, clan affiliation, for he now had links to the Upper Settlements through Great Tellico and Tannasse, to the Middle Settlements through Nequassee, and to the Lower Settlements through Ookunny and Keowee.

Taking all these points together, we find that an incident reported by an English traveler as a strong claim to European action based on a spatial and political understanding of control and domination is in fact a powerful display of Cherokee authority. Cummings journey graphically details the contrasting imaginings that the British and Cherokee had of each other. Each group's action and reaction were driven by differing spatial and political understandings, with each side comprehending the events in ways that made sense to themselves while simultaneously being unable to understand the other's reality.

---

[1] "Historical Relations of the Facts Delivered by Ludovick Grant, Indian Trader, to His Excellency the Governor of South Carolina," *South Carolina Historical and Genealogical Magazine*, 10 (1909), 56

[2] Gary B. Nash "The Image of the Indian in the Southern Colonial Mind," *William and Mary Quarterly,* 3rd Series, 29. 2 (Apr., 1972), 204.

[3] For details of Native American visits to Europe see Harald Prins, "To the land of the Mistigoches: American Indian Traveling to Europe in the Age of Exploration," *American Indian Culture and Research Journal.* 17.1 (1993); Carolyn Thomas Foreman, *Indians*

Abroad, 1493-1938, (Norman, Oklahoma, 1943); Christian F. Feest, ed., *Indians and Europe: An Interdisciplinary Collection of Essays* (Lincoln, Nebraska, 1999).

[4] Tore Frängsmyr, J.L. Heilbron, and Robin E. Rider., eds., *The Quantifying Spirit in the Eighteenth Century* (Berkeley, 1990).

[5] In 1703 at the tender age of 12 Cumming was granted a Captain's commission in the Earl of Mar regiment by Queen Anne and ten years later gained a Doctorate of Law from the University of Aberdeen. Over the next few years Cumming was to lead a company against the Jacobites in the 1715 uprising, turn down the Governorship of Bermuda, and in 1725, after his fathers death, became the second Baronet of Culter. As the decade drew to a close Cumming's eclectic career included a failed attempt to enter Scottish politics and admission to the Royal Society of London for Improving Natural Knowledge. Biographical details from The Scottish Tartans Museum Web site: http://www.scottishtartans.org/cuming.html Feb. 22, 2006.

[6] For details regarding the impetus for his trip see Daniel Lyons, *The Environs of London: Being an Historical Account of the Towns Villages and Hamlets, within Twelve Miles of that Capital: Interspersed with Biographical Anecdotes*, v.4 (London, 1795-96), 20

[7] Verner Winslow Crane, *The Southern Frontier, 1670-1732* (Ann Arbor, 1956), 277. Cumming spent a total of five months in the southeast arriving on December 6, 1729 and leaving on May 4, 1730. However, he was only among the Cherokee between March 13, and April 13, 1730.

[8] *The Daily Journal*, Number 3037, (Wednesday, Sep. 30, 1730) and Number 3044, (Thursday, Oct. 8 1730): The Journal was reprinted in *The Historical Register of 1731*; Note: Cumming refers to himself in the third person throughout the text.

[9] There is considerable overlap between the two reports; however, the details left out of the first "claim to fame" piece that appear in the daily journal enable us to gain a better understanding of the events that unfolded.

[10] See Herbert Eugene Bolton, ed., *Arrendondo's Historical Proof of Spain's Title to Georgia* (Berkeley, 1925).

[11] For the claim of Cumming's official role see: J.H. Stocqueler, *A Familiar History of the United States of America, From the Date of the Earliest Settlements Down to the Present Time* (London, 1865), p. 95; For his denial with regard to any official sanction of his trip see *The Daily Journal*, Thursday, Oct. 8, 1730.

[12] William O. Steele, *The Cherokee Crown of Tannassey*, (Charlotte, N.C., 1977), xiii.

[13] "Extract of a Letter form South Carolina, June 12", *The ECCHO: or Edinburgh Weekly Journal*, 89 (Wednesday Sept. 16, 1730); see also *The Grub Street Journal*, 36 (Thursday Sept. 10, 1730).

[14] Cumming returned to England shortly before payment of his loans became due. After his departure, in an attempt to recoup their money, several members of the Charles Town community broke into his treasury. All they found inside were "some empty Boxes, old Iron and other Rubbish" and it was "computed" that he carried off "no less than £15000 sterling" a substantial amount for the period. "Extract of a Letter form South Carolina"; Cumming's Manuscript, Ayer MS 204, Newberry Library, Chicago, 21.

[15] *The Daily Journal*, Thursday, Oct. 8, 1730.
[16] *The Daily Journal*, Thursday, Oct. 8, 1730; The cave is marked on George Hunters map to the south side of the Santee River slightly below the Township of Amelia.
[17] The Cherokee Path, mapped in 1730 by George Hunter, the Surveyor-General of the Province of South Carolina, ran for 130 miles from Charles Town and passed though the Cherokee Lower and Middle Towns before terminating at Settico in the Overhill Towns.
[18] *The Daily Journal*, Wednesday, Sep. 30, 1730.
[19] Veteran trader Ludovick Grant, who was in Keowee that evening, left a report corroborating the events that occurred in the Town House. See "Historical Relation of the Facts Delivered by Ludovick Grant, Indian Trader to His Excellency the Governor of South Carolina", *South Carolina Historical and Genealogical Magazine*, X (1909), 54-68.
[20] The named Europeans were Joseph Cooper, Ludovick Grant, Joseph Barker, Gregory Haines, Daniel Jenkinson, Thomas Goodale, William Cooper, William Hatton, and John Biles.
[21] *The Daily Journal*, Thursday, Oct. 8, 1730.
[22] Grant, "Historical relation," 56.
[23] *The Daily Journal*, Thursday, Oct. 8, 1730.
[24] James Adair, *History of the American Indians* (New York, 1974), 460.
[25] Long-term English traders also recognized this need for full national endorsement. While discussing a failed transfer of land from the Cherokee, Grant was to suggest that, the reason for failure was "because of the head men were not present." Grant, "Historical relation," 64; for a discussion of the "pragmatic" attitude by the Cherokee to external threats see Robert K. Thomas, *Cherokee Values and World View*, (Chapel Hill, N.C., 1958), 14.
[26] Randolph Boehm, ed., "Journal of the Commissioner for Indian Affairs on his Journey to the Cherokees and his Proceedings there", *Records of the British Colonial Office, Class 5 Part 1: Westward Expansion 1700-1783*, v.12, (Frederick, MD., 1972).
[27] This harmony ethic has continued to be an important part of Cherokee life. See John J. Gulick, "The self-corrective service and persistence in Conservation Eastern Cherokee culture", *Research Previews*, (1959).
[28] Raymond D. Fogelson "Cherokee Notions of Power", in Raymond D. Fogelson and Richard N. Adams, eds., *The Anthropology of Power: Ethnographic Studies from Asia, Oceania, and the New World* (New York, 1977), 189-90.
[29] Ibid.
[30] "Account of the Cherrokee Indians, and of Sir Alexander Cuming's Journey Amongst Them", *The Historical Register*, 16, no. 61 (1731), 3.
[31] Alexander Longe, "Small Postscript on the ways and manners of the Indians called Cherokees" *Southern Indian Studies,* 21, (October 1969), 39.
[32] *The Daily Journal*, Wednesday, Sep. 30, 1730 and Thursday, October 8 1730.
[33] *The Daily Journal*, Thursday, Oct. 8, 1730.
[34] *The Daily Journal*, Thursday, Oct. 8, 1730.

[35] It must be remembered that the translator, or linguist, in these negotiations was Ludovick Grant, a trader based in Great Tellico who therefore had reason to promote the importance of the leader of his Town.
[36] *The Daily Journal*, Wednesday, Sep. 30, 1730.
[37] *ibid.*
[38] *ibid.*
[39] *ibid.*
[40] British and Cherokee spatial understandings are examined in my doctoral dissertation *Space the Final Frontier: Spatial Understandings in the eighteenth century American Southeast*, (University of California, Riverside, 2006).
[41] *The Daily Journal*, Wednesday, Sep. 30, 1730.
[42] Colonel George Chicken was father to the George Chicken who later accompanied Cumming.
[43] "Colonel Chicken's Journal to the Cherokee, 1725" Newton D. Mereness, *Travels in the American Colonies*, (New York, 1916), 101.
[44] Ludovick Grant described Cumming's "crown" as a "cap" worn by beloved men which "resembles a wig and is made of Possum's Hair Dyed Red or Yellow." Grant "Historical Relation," 57.
[45] *The Daily Journal*, Wednesday, Sep. 30, 1730.
[46] *The Daily Journal*, Thursday, Oct. 8, 1730.
[47] *Ibid.*
[48] Grant "Historical Relation," 57.
[49] *The Daily Journal*, Wednesday, Sep. 30, 1730.
[50] Cumming's Manuscript, Ayer MS 204, Newberry Library, Chicago, 22 and 26.
[51] Cumming reports that the "King, who had been just then made at Ookunny" was the same person as the "King of Keowee", *The Daily Journal*, Wednesday, Sep. 30, 1730.

CHAPTER FOUR

SYMBIOTIC STRENGTH: AN EIGHTEENTH-CENTURY VIEW OF THE BRITISH-AMERICAN RELATIONSHIP

JOHN BARRINGTON

Of all the Britons who have participated in discussions of America over the centuries, no group has had more at stake than the pre-Revolutionary American colonists. These colonists took great pride in being British, and at the same time saw America as their home. Every British commentary on America made an implicit judgment on the sustainability of the colonists' identity by raising the question: could Britons in America remain fully British? Given the importance of this issue to British-Americans, it is unsurprising to find that they not only eagerly read discussions of America produced in the mother country, but also reformulated these discussions to meet their own needs. The present chapter will explore the discourse on British-American relations developed by one particular set of colonists, the literate whites of Charleston and Low Country South Carolina. This discourse presented a cogent solution to the conundrum of how residents of a land so different from Great Britain could not only remain fully British, but even be essential to the survival of all that was admirable in the mother country.

Colonists participated in discussions about America and the British-American relationship in a number of forums—letters, diaries, sermons, the meetings and official papers of the provincial government, and innumerable private conversations and thoughts that were never recorded. No forum for discussing this relationship was held so regularly, or embraced so many participants, as the newspaper. The only newspaper printed in South Carolina before 1758 was the *South-Carolina Gazette*, one of the longest-running periodicals in eighteenth-century British America, with an almost unbroken existence from 1732 until 1777. The present chapter examines in detail this newspaper's discussion during the 1750s of America's place in the British Empire. By examining this decade, it is possible to discover, in its most fully-evolved form, how literate South Carolinians viewed the British-American relationship before the disruption of that relationship by the revolutionary crisis.

The *Gazette*, in common with most newspapers of the eighteenth century, consisted of a mixture of locally generated material—advertising, local news, letters from real and invented readers, and editorial comment—combined with a great deal of information copied from newspapers printed in other cities of the Empire. Peter Timothy, the *Gazette's* editor, thus created a worldview for his readers by selecting, reorganizing, and supplementing a broader discussion being conducted in Britain and other colonies. Readers shaped the newspaper's contents both by their letters and other contributions, and by purchasing the periodical and thus demonstrating an interest in its contents.[1] The information and commentary published in the *Gazette* simultaneously voiced the particular viewpoint of a single province and manifested a wider, transatlantic discourse about America's place in the Empire and the world.

Three points stand out from the *Gazette's* discussion of British America in the 1750s. First, the newspaper consistently assumed that the colonists were fully British, sharing important privileges and distinguishing characteristics with Britons at home. The degree to which these assumptions permeate the contents of the *Gazette* is unsurprising, given the large volume of scholarship that has uncovered, in many different contexts, the intense loyalty towards Great Britain felt by most Americans right up to the eve of independence.[2] In this respect, the present study simply adds one more piece of evidence to a well-established understanding of colonial identity.

The second and third parts of this chapter, on the other hand, discuss much less widely appreciated aspects of how colonists understood America's connection to the mother country. The second part explores the *Gazette's* arguments about the vital role of South Carolina and other colonies in Britain's mercantilist system. Far from resenting trade regulations and restrictions, the *Gazette* regarded the colonies' supply of exotic raw materials as a vital patriotic duty, since such materials made the Empire economically more independent of its enemies and supported the defense of British religious and political liberty against France and Spain.[3] The third part of this chapter shows how the *Gazette* was able to absorb America's distinctive populations, including native Americans and Africans, into the overarching narrative of Britain's struggle against its Catholic foes. Readers of this newspaper clearly felt they were fully British not only because of what they shared with the mother country, but also because of what was unique about America.

Shared Characteristics: Colonists as Britons

The *Gazette* stressed several characteristics that South Carolinians shared with Britons at home, including nationality, loyalty to the king, and the enjoyment of distinctive religious and political liberties. These common traits have been widely explored by scholars and need not be labored here. A few illustrations will demonstrate how the *Gazette* applied these British characteristics to South Carolinians and to other colonists.

The *Gazette* repeatedly linked Britons in America and in the mother country by describing the colonists simply as "British" or "English". Despite the differences of meaning between these two terms, the *Gazette* in practice used them interchangeably to describe individuals from any or all of the provinces of the British Empire. For example, one issue referred to ships captured by the Spaniards and held at Havana as "English" ships, although they came from New York, Barbados, and Ireland. The same term was used to describe ships in the Bay of Honduras that originated from a variety of imperial ports, including Charleston.[4] In so far as there was any difference in the usage of these terms, "British" tended to be used in emotive contexts, where the loyalty of individuals towards the empire was being stressed, such as in the eulogy for Peter Mercier, a South Carolinian killed during the French attack on Washington's force at Fort Necessity in July, 1754.[5] "English" was more often used neutrally, simply to mean imperial subjects.

Aside from these national labels, the *Gazette* stressed colonists' British or English identity by emphasizing Americans' loyalty to the monarch. One regular manifestation of this loyalty was the celebration of royal anniversaries. For example, South Carolina chose the anniversary of George II's accession to the throne (June 22) as the date on which to lay the foundation stone for the colonial government's new quarters.[6] A more frequently observed anniversary than that of the king's succession was his birthday, November 10. The birthday festivities in Charleston in 1754 were typical of what took place on these occasions. Cannon were fired, the Governor and Council reviewed the colony's Independent Company, "and in the Evening, His Excellency gave an elegant Entertainment at Gordon's and a Ball to the Ladies which lasted 'till the Morning." South Carolinians could read about these and similar festivities in other cities of the empire, such as London and Philadelphia, so that readers could imagine British Americans and Britons in the mother country simultaneously expressing affection for the king.[7]

George II helped to connect not only his American subjects, but also his American lands to Great Britain. Frequently, the territories of the British Empire in America were described as the king's land, counteracting the geographical separation of the New World from the Old, and bringing a conceptual solidity to a large, ever-changing, and often vaguely-defined territorial expanse.[8] One evocative way in which the *Gazette* stressed the association of imperial territory in America with the king was in a report of a ritual enacted on New England's northeastern frontier. When Massachusetts sent a force up the Kennebec River to expel French missionaries and to claim the land for the British Empire, commander John Winslow formally drank the king's health at the new border.[9] The concept of the king's land was flexible enough to make every newly acquired or claimed territory in America, however distant or wild it might be, into an immediate extension of Great Britain.

The essential benefit of being a subject of the British king and of living on his land was that George II was a constitutional monarch, who ruled for the sake of his subjects' interests, rather than for his own. The *Gazette* frequently reported news that reminded readers of the benevolent government and essential liberties guaranteed by the British constitution to Britons on both sides of the Atlantic, liberties that included religious freedom (for Protestants), protection of private property, freedom of the press, and equality before the law. For example, the *Gazette* printed an address by the House of Lords that spelled out the various aspects of British liberty that the King was expected to foster. "The honour and security of the nation, both at home and abroad, the maintenance of our religion and liberty, the protection and extension of our commerce, and every branch of national happiness are the objects of your royal care, wisely and steadily exerted for the common good of your people."[10] Speeches by royal governors to colonial legislatures made it clear that the same liberties were shared by the American colonists. Governor Robert Dinwiddie of Virginia exhorted the Virginia legislature to be mindful of "the liberties, the properties, and the pure religion" that they enjoyed, and both legislative houses asserted their determination to protect these blessings.[11] South Carolina's Governor James Glen called George II a king who has "never invaded the Privilege of the meanest of His Subjects" and he asserted that it is the "peculiar Excellency of His Majesty's Government, that all His Subjects may be happy if they will." Indeed, Glen boasted, "We enjoy the happiest and most perfect Frame of Government in the World: it is the Envy of all Nations: the Language of all Nations is, Who would not be a *Briton*?" The South Carolina Commons House of Assembly agreed that "We are truly Sensible of the Happiness peculiar to His Majesty's Subjects, remote as we are from His Majesty's Royal Presence; and we are assured, this Happiness was never more compleat than it is at present, under the benign Influence of His Majesty's Government...[so that] we shall look upon His Majesty's Interest and our own as inseparably united."[12]

British liberties were celebrated by ordinary folk as well as by those in government: an account of an election at Ipswich (England), in which the popular naval hero, Admiral Edward Vernon, was returned to Parliament, described 2000 supporters marching in a procession, "two and two, all with blue cockades in their hats, and the following motto in silver letters, Christianity, Liberty, and Loyalty."[13] In an electoral contest at Henley-on-Thames, victorious candidates led a procession that carried flags bearing the slogans "Liberty and Loyalty" and "the protestant religion and succession."[14] Members of the public also used the courts to defend a variety of rights. The *Gazette* discussed in detail the trial and acquittal of a Mr. Owen, a London bookseller and printer, who was pronounced "Not Guilty" in a seditious libel trial, "to the entire satisfaction of a numerous and crowded audience, who were unanimously delighted to find the liberty of the press (so intimately connected with the liberty of the subject) so justly asserted by this equitable verdict."[15] The *Gazette* also printed the story of two "patriot coblers" who demonstrated the rights of the lower classes, one by suing the mayor of an English borough to force him to hold an election, and one who reopened a right of way across an aristocrat's private park. The *Gazette* concluded these reports with a comment by the Duke of Devonshire, who said "the liberty of a cobler ought to be as much regarded as that of anybody else; that is the happiness of our constitution."[16]

The *Gazette* included no accounts of humble colonists celebrating or protecting their liberties in election-day marches or legal battles. However, the newspaper made it clear that ordinary colonists cared as much about liberty as did ordinary Britons at home: the chief manifestations of this British-American attachment to liberty consisted of military service and of contributions to the imperial economy and fiscal system. Britons in America were equally English, equally loyal subjects of the monarch and possessors of the liberties enjoyed by the British constitution. Yet British Americans were able to serve the monarch and nation in a distinctive way, by making available the New World's unique products.

Explaining Away Differences: Mercantilism and Imperial Interdependence

One of the most frequently reiterated themes in the *Gazette* was the challenge faced by the British nation from foreign, Catholic powers, especially France and Spain, who were seeking to overrun all or part of the Empire. Most of the news and much of the editorial commentary included in the *Gazette* during the 1750s related to this Catholic menace in some way or another. The newspaper warned readers of the disturbing possibility that British liberty would be extinguished unless Britons on both sides of the Atlantic did their utmost to avert the threat.

The most serious danger posed by Britain's Catholic enemies was the possibility that they might reinstate the exiled Stuart dynasty, and thereby destroy the freedom enjoyed in the British Empire since the Glorious Revolution. The *Gazette* took this possibility seriously, and reported Jacobite activity in detail. News of the Pope's involvement with the Stuarts was one area of concern: the *Gazette* told readers that the Pope had patched up a quarrel between the Pretender and his second son, the Cardinal Archbishop of York, and had appointed another Cardinal as a protector of Scotland.[17] The *Gazette* also kept readers informed about the gathering of weapons and the return of former rebels to Scotland in 1753.[18] France was suspected of planning to assist a new Stuart invasion: the newspaper reported military preparations at Dunkirk, the most likely base for an invasion of England, the movements of French ships off the Scottish coast, and the presence of Jacobites in France.[19] The Pretender's eldest son, who had personally led the 1745 invasion, was also the subject of discussion: his possible marriage, his possible homosexuality, and his rumored conversion to Protestantism all attracted notice.[20] While the Papacy and France were the chief backers of the Stuarts, Spain and Spain's allies in Italy (especially Naples) were also assumed to be secret supporters of the exiled dynasty.[21] In addition to fears of current Jacobite plots, the *Gazette* reminded readers of the events of the Glorious Revolution. Among other detailed references to this foundational event was the report of the death of a minor customs official who had accidentally found James II hiding on a ship at the time of William of Orange's invasion.[22] The *Gazette* also described the "usual" celebrations in Charleston of November 5, the "Anniversary of our happy Deliverance from a most horrid *Popish* Plot [the Gunpowder Plot of 1605] and of the glorious Revolution by the landing of King William in England."[23] In addition, every reference to the "Protestant Succession" during rhetorical celebrations of British liberty reminded readers of the constitutional blessings they enjoyed thanks to the Stuarts' expulsion.

While the Catholic Powers' support for the Stuarts was the most serious cause for concern, the struggle between Britain and its enemies manifested itself in many different ways across much of the globe. The *Gazette* kept South Carolinians regularly updated about the details of diplomatic, military, and naval developments in an area bounded by the Mediterranean, the Ottoman Empire, Russia, Scandinavia, the British Isles, Greenland, Britain's North American territories, the West Indies, the Bays of Campeche and Honduras, and the Mosquito Coast. Reports from Persia and India were more sporadic, but frequent enough to give readers some idea of what was happening in those regions. Occasional mentions of West Africa, the Rio de la Plata, the Dutch East Indies, and China provided glimpses of more distant regions. The *Gazette* made it clear that the global context was all-important for understanding the threat to British liberty. When, at the end of 1753, the editor summarized the nine most important disputes between the European states, his list embraced a region that stretched from the Ohio Valley to India.[24] The American colonists' liberties were thus intimately connected to European and Asian affairs. For example, the *Gazette* reminded readers that civil war in Persia embroiled Russia and the Ottoman Empire, freeing France to take action against the British Empire, including British America.[25] South Carolinians should learn from the news, urged one contributor: summing up French activities in Sweden, Dunkirk, the West Indies, Cape Breton, and the Ohio, this writer concluded that South Carolina needed to invest heavily in fortifications.[26] The *Gazette* thus presented British America as anything but a world unto itself: the American colonies were an integral part of a globally-engaged empire.

British subjects in America did more than simply celebrate liberty and worry about French and Spanish aggression. The *Gazette* kept its readers fully aware of the fact that defense of the Empire's liberties depended above all on the Empire's resources, in Britain and in Britain's colonies. As Governor Glen phrased it, in an address to the South Carolina legislature, military strength was chiefly the product of "Numbers [of people] and Wealth, Men and Money being now the Nerves and Sinews of War in all Countries."[27] Money for defense came from trade, above all, and so it was vital to protect every British colony and every British trade route, not just for the sake of prosperity, but in order to underpin the costs of defending the British constitution and British liberty. The *Gazette* reminded its readers that the loss of trade to foreign powers was the first step towards national disaster: after foreign powers have "plucked all the feathers, [they] will devour the entire country like a naked bird."[28] This linkage of resources with armed might constituted a worldview that can usefully be labeled "mercantilist", even though that particular term was not used in the *Gazette*.

The newspaper illustrated this mercantilist understanding of national strength in its detailed reporting of the means used by Britain and the other European powers to build up their economies and to attract immigrants in order to strengthen their armed forces. A report from London informed readers that there were now 45 British ships engaged in the Greenland fishing and whaling trade, which was "of greatest importance to the welfare, as well as the naval force, of this kingdom."[29] Similarly, the North Sea herring fishery was vital: it not only employed 20,000 people and relieved poverty, but, by training sailors, was "an addition to our naval strength."[30] Spain, too, was engaged in building up its economy, in order to build up its armed forces: the government had compiled a book of instructions for the woolen weavers and had passed laws to encourage that industry, while it had also founded schools of navigation and shipbuilding, and was busy fortifying its major ports. All of this activity suggested that Spain seemed "to aim at the title of a maritime power."[31] Attracting immigrants was part of the process of building economies: the

Gazette reported that Protestant refugees from France were settling in Switzerland and Prussia, the latter providing incentives to bring in immigrants skilled in silk and velvet manufacturing.[32] These and other reports made it clear that a skilled workforce, shipping, industry, and armed power were intimately connected. Every detail that appeared in the *Gazette* about commercial treaties, economic incentives, the build-up of armies and navies, and diplomatic negotiations were part of one long narrative about the ever-shifting balance of power in Europe.

America played a vital and distinctive role in this world-wide struggle to create wealth and maritime trade. One of America's most important products was sugar, and the *Gazette* had strong opinions about this commodity. One concern was that the French Caribbean islands were producing sugar more cheaply than the British islands, leading to pressure on the British government to allow the importation of the French product. "Friends of our Colonies" in Britain were resisting these imports, but the colonies had to take action as well: Jamaica, it was suggested, should break up more land for sugar production. That island could also produce the luxury wood that the British were having to obtain from the Bay of Campeche, in the face of fierce Spanish opposition.[33] Back in Europe, French fishing off the British coast should be prevented, since the French used the catch as a cheap way of feeding the slaves on its American sugar islands.[34] Discussions such as these demonstrated that mother countries and colonies were closely connected in the economic competition between the major powers.

Along with the rest of British America, South Carolina was intimately involved in the British mercantilist system, according to the *Gazette*. The newspaper revealed South Carolina's commercial links to the British world in a variety of ways. For example, advertising, which occupied a significant proportion of each issue, illustrated the role of imperial commerce in the colony by alerting readers to the many different types of merchandise available for purchase. Another incidental illustration of the close link between the colony and imperial trade came in the wake of the devastating hurricane that struck Charleston on September 15, 1752. The list of ships in port that had been damaged by the storm included vessels connecting Charleston with New York, Rhode Island, Hull, Bristol, Falmouth, Liverpool, Jamaica, Halifax, Cape Fear, and Barbados. In addition, a British warship, the *Mermaid,* had been driven aground by the tempest. Imperial commerce and defense turned this item of local news into an illustration of South Carolina's place in the imperial community: the disaster affected Britons from all over the Atlantic empire.[35]

The *Gazette* not only revealed the existing commercial links between South Carolina and the rest of the Empire but also discussed ways of intensifying those ties. One writer argued that such discussions were, in fact, a major reason for the newspaper's existence: contributors to the *Gazette* should find ways of using "every interval of peace, for future security in war, extending [the colony's] commerce, peopling its frontiers, and providing for their protection and prosperity, etc."[36] Indigo, a fairly new product from the colony, attracted a lot of comment in the *Gazette's* pages. Governor Glen praised the product for its value to both the local economy and to the empire.[37] The value to the Empire was clear, since Admiral Vernon's "Anti-Gallican" (Anti-French) society in Britain offered a prize to the South Carolinian who produced the most indigo in 1754.[38] The *Gazette* reported on a hurricane that had devastated the indigo crop on Espanola, making South Carolina's product more valuable in European markets.[39] It also carried advertisements for "Guatemalan Indico Seed" that would allow more producers to grow the crop.[40] The South Carolina Indigo Society was founded to spread knowledge of indigo cultivation to its members.[41] Of course, the negative aspect of indigo's value was that it would inevitably attract foreign invaders: a contributor to the *Gazette* warned that the colony had to fortify itself in order to ward off France's desire for the indigo plantations.[42] Indigo provided many opportunities for *Gazette* readers to consider the integral role that their colony played in the Empire's mercantilist system: individual South Carolinians, through their engagement in a profitable economic activity, were deeply involved in the contest for domination among the European powers.

The present and future success of indigo prompted other suggestions from *Gazette* readers about how their colony could benefit the British Empire, at the same time as it built prosperity for itself. One contributor argued that South Carolina could encourage skilled immigrants from Sweden and Russia to teach British immigrants how to grow hemp and how to make pot-ash and tar. Such products would reduce the British Empire's dependence on foreign nations for products vital in shipbuilding and in making gunpowder. There were pressing reasons for avoiding such dependence: France was very influential in Sweden, and might persuade the Swedes to stop trading with Britain and to block British trade with Russia.[43] South Carolina could thus play an economically strategic role in the Empire: such possibilities were further reasons for readers to study the news of distant places in the *Gazette.* The sense that the colony's identity was intimately bound up with imperial trade was demonstrated by a satirical suggestion that South Carolina could export its surplus of attractive young, white women to Britain, since South Carolina's white males were too interested in their female slaves to need female company of their own race. The use of a mercantile metaphor in this critique of interracial sex demonstrates the pervasiveness of ideas about transatlantic commerce.[44]

The value of American products in Britain's mercantilist system meant that America, like other parts of the British Empire, was under constant threat of attack from France and Spain. The *Gazette* informed readers of these attacks, not just because they were of local interest to British Americans, but because they were part of the broader power struggle between Britain and its Catholic enemies. One constant problem came from the activities of Spain's *Guarda Costa*, which stopped British (including colonial) shipping in the West Indies, and searched the ships for evidence of trade goods or money smuggled out of Spanish America. The *Gazette* supplied detailed information about the most outrageous of these seizures, as well as updates about British negotiations in Madrid that were aimed at curtailing the *Guarda Costa's* activities.[45] The *Gazette* also reported frequently about Spain's efforts to drive the British (including South Carolinians and other colonists) from logging settlements in the Bay of Campeche, the Bay of Honduras, and the Mosquito Coast.[46] Less frequently mentioned, but still disturbing to South Carolinians in the wake of the Stono Rebellion of 1739, was Spain's encouragement to African slaves in South Carolina and to Indians allied with the British in South Carolina and Georgia to leave British territory and to settle near St. Augustine, Florida.[47]

France was an even greater threat than Spain. That nation was building forts in the Ohio Valley, arresting British traders who ventured into that area, and encouraging Indian allies to attack frontier settlements and tribes friendly to Britain, from Pennsylvania down as far as Georgia.[48] These French incursions in the Ohio Valley were "no new or partial scheme of the French, merely for the sake of trade, or a settlement in the lands, but a thing long ago concerted, and only part of a general plan for rendering themselves masters of North-America ... this plan has been concerted, laid before the court of France, and met with its highest approbation in the year 1689."[49] Further north, the French in the St. Lawrence Valley were encouraging similar Indian raids on New York and New England, while the French forts in Acadia and Louisbourg were bases for provoking Indians and the "Neutral French" to make trouble for the British in Nova Scotia.[50] In the West Indies, the French moved in on the Windward Islands, threatening to close a vital communication route between the British Caribbean and British North America.[51] A threat to American commerce rather than to American territory was posed by French involvement in Corsica, where France had Spain's blessing or possibly even Spanish aid. If France were to succeed in crushing the Corsican rebels, Corsica would become a naval base from which French fleets could menace British trade in the Mediterranean, including South Carolina's export trade in rice.[52] As these specific conflicts escalated into a general, though still undeclared, war with France over the course of 1754 and 1755, the French began to threaten British (including colonial) shipping everywhere in the West Indies and off the coast of Africa.[53] The *Gazette* made it clear that the Catholic Powers would take advantage of every circumstance to gain trade and territory at Britain's expense: for example, when Charleston and its defenses were devastated by the 1752 hurricane, the government's immediate action was to call an emergency session of the Assembly to pass measures to "make His Majesty's Subjects as secure as our present Circumstances will permit, against all Attempts from without."[54] 1752 was a time of official peace between Britain, France, and Spain, yet even the weather became a potential vehicle for an invasion by Britain's Catholic enemies. This assumption was not unwarranted for those who believed, as one writer did, that the French, at least, "have ever been aspiring at universal Monarchy."[55] The unending efforts of France and Spain to conquer Britain's American colonies demonstrated, more powerfully than anything else, the immense value of those colonies to Great Britain and to British liberty.

Explaining Away Differences: America's Distinctive Populations

By drawing on mercantilist ideas, the *Gazette* was able to transform the huge differences between Great Britain and America's tropical and subtropical regions into a powerful bond: only through the economic and defensive efforts of both areas could British liberty be defended from the Catholic powers. The same struggle between British Protestantism and "popery" enabled the *Gazette* to convert another major difference between the New World and the mother country into a common tie: British America was home to a mixture of ethnic groups that made it increasingly distinct from Britain, but those groups were all subsumed into the fundamental Catholic-Protestant struggle. Three categories of ethnic groups were construed in this way: non-British migrants from Europe to British North America, native Americans, and enslaved Africans.

Although Protestant migrants from the British Isles and their descendents still constituted the majority of white settlers in British North America, immigration and conquest had created significant minorities of other white ethnicities, particularly French and German. For the editor and readers of the *Gazette*, however, religious differences overrode ethnic ones, making America not into a "melting pot", but into a mirror of the mother country, where a Protestant majority watched carefully for revolts and other plots from Catholic minorities. One important group of non-British, Protestant migrants in South Carolina were the French Huguenots, including the parents of Peter Timothy, the editor of the *Gazette*. Not surprisingly, Timothy included a great deal of information in his newspaper about religious issues in France, reflecting his and some of his readers' ancestry. This news from outside the Empire, however, was easily assimilated to the mainstream concerns of Protestant Britons.

The most closely covered French religious news in the *Gazette* of the 1750s was the persecution of the Jansenists. Although the Jansenists were in fact a group within the Catholic Church, the South Carolina paper represented their beliefs as being similar to Protestantism: Jansenists were being persecuted for believing that the laity should read the Bible for themselves, and make it central to their worship.[56] The religious controversy spilled over into a political battle: the Parlements, or law courts of France, supported the right of the Jansenists to worship as they wished, provoking the king to disband or to threaten the Parlements, until France seemed poised on the verge of civil war. From October 3, 1752 until the spring of 1754, the *Gazette* covered the controversy in almost every issue. The *Gazette* frequently pointed out the political implications of the Catholic Church's persecution of the Jansenists. The Paris Parlement warned the king about "an empire which is rising in the midst of your state; an arbitrary empire, which acknowledges no laws, no sovereign, no magistrates, and to which religion is only a pretext: the authority of the prince an instrument which it presumes to use or lay aside according to its interests...."[57] The influence of the clergy was infecting the French monarchy with false principles, drawn from the Catholic religion, particularly the idea that "there is, or ought to be, *an infallible Head* in the State, as well as in the Church; which can hardly be acknowledged by Men that have either Hearts or Heads."[58] The clergy's desire to persecute heretics was overturning what little liberty was left in France, and British subjects were invited by the *Gazette* to regard the Parlements as fighting for civil and religious liberties similar to those that Britons themselves so prized.[59] While Timothy's decision to cover the controversy in such detail was very likely a personal decision, motivated by his interest in his parents' country of origin, he could not have continued to sell papers (and gain the advertising revenue that depended on a robust circulation) if his readers—Britons as well as Huguenots—had been unwilling to read this material. Timothy's readers were interested in the Jansenist controversy because they could align themselves sympathetically with the victims of Catholic tyranny, and could imagine what would happen if Catholicism were to gain power in the British Empire as a result of French, Jacobite, or Spanish victory in war. An overarching Protestant identity thus enabled the concerns of this Huguenot editor to be merged with the broader British struggle against Catholic foes.

Just as the concerns of French Protestant migrants could be linked to the concerns of British Protestants, so the behavior of French, German, and other Catholics in the colonies could be equated with the plotting of Irish, Scots, and English Catholics at home. One traditional anti-Catholic stereotype in the English-speaking world was that Catholics, being determined to impose their faith on their Protestant neighbors at all costs, would take and then break oaths of loyalty, believing that their religion absolved them of the duty to keep agreements with heretics. Catholic populations would rebel on any pretext, and lone Catholics were consummate conspirators.[60] The *Gazette* provided evidence that Britons living in America faced the same threats from internal Catholics as did Britons in the mother country. Acadia was a particular problem. At the end of 1752, the usual restrictions on allowing Catholic immigrants into America were relaxed, so as to allow some impoverished German Catholics into Nova Scotia. Within two years, religious ties had proved stronger than the gratitude which these Germans ought to have felt towards their British patrons: the French Acadians convinced the Germans that they were ill-treated, and stirred them into a revolt.[61] In 1755, when the *Gazette* covered the round-up and dispersal of the Acadians, it justified this harsh measure as a preemptive strike against blatant anti-British and anti-Protestant conspirators.[62] Pennsylvania, the only colony except Nova Scotia where Catholics were permitted to build churches, also provided evidence of Catholic ingratitude and conspiracy. The *Gazette* reported that "a large quantity of gunpowder has lately been discovered in the Romish chapel at Philadelphia, where it was secreted, we suppose, by that people, 'till they should find time and opportunity to make it serviceable, in reducing the Hereticks to the passive obedience and non-resistance of the true Catholic Faith."[63] The *Gazette* also reported that two colonial companies at Saratoga, on the northern frontier against French Canada, discovered that the cartouches with which they loaded their guns contained no bullets, rendering them useless in battle. It was then discovered that the man who had made up these cartouches was an Irishman who had formerly served the French in Canada. Although he had been living in Connecticut for two years, he still, it seemed, retained a primary loyalty to his co-religionists.[64] These reports were interspersed with many tales of continued Catholic conspiracy in the British Isles, emphasizing the degree to which the colonists shared the same insecurities as the mother country. Even though migration had created a different ethnic mix of Catholics in America than in the British Isles, these French and German Catholics appeared to be just as untrustworthy as Catholics back home.

It was, perhaps, not surprising that the *Gazette* was able to equate the behavior of the Catholic inhabitants of North America with Catholic minorities in the British Isles. More striking is the newspaper's regular attempt to portray the bloodthirsty hostility of native Americans towards the colonists as another manifestation of "popery's" relentless cruelty towards Protestants. The *Gazette* did not claim that large numbers of Indians had become Catholics, but rather focused on the way in which Catholic agents were able to stir up the worst instincts of the natives: the trope of "popish" cruelty, well established in anti-Catholic literature, was extended to and blended with the stereotype of Indian savagery.[65] Readers discovered that Spanish Jesuits were inciting the Indians of the Mosquito Coast to murder all the British settlers there, just as French priests and laity provoked the natives on the North American continent to massacre innocent Britons on the frontier.[66] A Monsieur Morin and his son, working with the Indians to murder British traders in the Ohio, had a tobacco pouch made out of the skin of a Boston ship's captain. The rest of the captain's flesh had been eaten by Morin and by some Indians, whom he had incited to this act of cannibalism.[67] One of many descriptions of massacres along the frontier reported that "Nothing is to be seen but desolation and murder, heightened with every barbarous circumstance, and new instances of cruelty. They [the Indians], at the instigation of the French, burn all the plantations."[68] Another report claimed that papers taken from a French officer showed that the French, assisted by Indians, planned to seize Halifax, Nova Scotia, whose "inhabitants [were] all to be shut up in the church, and fire put to it, and the troops all to be put to the sword without quarter."[69] One group of "French" Indians actually refused to attack the settlement at Saratoga, New York, because too much "English" blood had already been shed.[70] French cruelty was highlighted, not only by Indian restraint, but by the humanity of the British: General Braddock, the *Gazette* reported, forbade his Indian allies from scalping their enemies.[71] These reports enabled the *Gazette* to make the native inhabitants of America, frequently an emblem of American distinctiveness, into a manifestation of the same Catholic foe faced by Britons in the mother country.

If native Americans made America ethnically distinct from Great Britain, so did African slavery, especially in the case of the southern mainland and Caribbean colonies. Many historians have stressed the fact that enslaved Africans constituted the majority of South Carolina's inhabitants from almost the start of the eighteenth century. These scholars have argued that this black majority made white South Carolinians feel increasingly insecure, and made South Carolina's society and worldview entirely distinct.[72] Evidence from the *South-Carolina Gazette*, however, points to different conclusions about how South Carolina's whites viewed slavery. First, it is clear that South Carolinians feared Catholic enemies far more than slave insurrections, and that slave insurrections were taken most seriously when they were part of the broader Catholic threat to British Protestantism. Second, mercantilist arguments enabled the *Gazette* to present slavery as an institution that contributed to colonial production and therefore to imperial defense, so that slavery, on balance, was actually a source of security against the Empire's Protestant foes. These arguments suggest that even widespread slavery could serve to reinforce the idea that Britain's American colonies were an integral part of the British world.

Of course, the *Gazette* did not take the dangers and inconveniences of slavery lightly. While there were no large-scale slave revolts or conspiracies in South Carolina during the 1750s, reports of revolts in other places must have reminded *Gazette* readers of what could happen in their own colony.[73] In response to that possibility, contributors to the *Gazette* regularly wrote about the need for careful patrols, for the enforcement of laws that ordered every ten slaves to be supervised by one white, and for preventing slaves from carrying weapons.[74] Slave revolts might occur not only on land, but at sea, where slaves occasionally took control of slaving vessels off the coast of Africa.[75] Slaves also committed crimes, like murder and theft.[76] Slaves on one occasion provoked trouble with native Americans by murdering an Indian in a drunken brawl.[77] Slaves, of course, did "steal themselves" by escaping, as advertisements regularly pointed out. The presence of slave women tempted white Carolinian men to indulge in interracial sex, an activity that the *Gazette* branded as grotesque and ridiculous.[78] Newly imported slaves could also bring disease into the colony.[79] It is easy to find evidence that slavery created many problems for whites, especially by selectively reading the *Gazette* to find examples of such problems.

If read in context, however, the insecurities created by slavery are outweighed by other factors. Most of the *Gazette's* reports suggest that white South Carolinians regarded their colony as part of a broad theatre of Great Power conflict, stretching from the Ohio to the Ganges. In the context of this theatre, the African slaves appeared isolated and friendless, while South Carolina's whites were intimately connected to a major world power. The *Gazette* reflected the whites' sense of their slaves' isolation by providing absolutely no information about political, diplomatic, or military affairs in Africa. Although some information about African regimes must have been available from the captains of slavers who called at Charleston, there were no reports about even coastal polities. Given the ability of South Carolinians to read about Russian trade policy on the Caspian Sea, French diplomacy at Delhi, and the numbers of each nation's ships fishing off the coast of Greenland, this lack of information about Africa is significant. Apart from indicating a lack of respect for African culture and civilization, the *Gazette's* silence about this topic conveyed a comforting message to its readers: Africans by themselves could launch only a local revolt against whites, since they were not part of any broad national community with the wealth to support sophisticated navies and armies. Whites faced by African revolt, on the other hand, could call on a large and powerful community of fellow nationals to help suppress the uprising. Slave revolt would be an unpleasant prospect, but there was no possibility of an ultimate African victory, if the Africans alone were taken into consideration.

Of course, given the state of virtual or actual war between the British Empire and its Catholic enemies, rebellious Africans would not be alone. France and Spain could provide the organization, the weapons, and the backing from a large, modern state, that slave populations needed for a successful revolt. White South Carolinians were most fearful of the black majority when it appeared to be yet another tentacle of the "popish" monster which, unlike the slaves, did have a real prospect of overrunning all or part of the British Empire on a permanent basis. *Gazette* readers were reminded, on a small scale, of the possible connection between the Catholic powers and the slaves whenever small groups of Africans escaped to find freedom—and to be converted to "popery" —in Spanish or French territory.[80] The *Gazette* also recorded fears of large-scale revolt by all the slaves in the colony in the wake of a Spanish or French invasion.[81] Conclusive evidence that South Carolinians feared Catholics more than their slaves emerges from the colony's immigration policy: slaves were far preferable to white, Catholic immigrants in the colony, since, as reports from across the empire showed, Catholics habitually rebelled or conspired.[82] For this reason, when South Carolinians talked of encouraging white immigration, it was always Protestant immigrants who were mentioned. When white Protestants weren't available, enslaved Africans were preferable to white Catholics.

The need for white Protestants or African slaves stemmed from the mercantilist assumptions that permeated the *Gazette's* pages. Immigrants, including enslaved immigrants, were needed to boost the wealth and the defense capability of the colony and the Empire. Since Britain's enemies used slaves to increase their colonial production, imperial wealth, and armed power, Britain needed slaves to keep the balance of wealth and power favorably tipped in its favor. Slave labour was often the only way of turning empty, cultivatable lands into viable settlements, and if Britons neglected to cultivate vacant lands in this way, France or Spain would move into the vacuum.[83] While slaves might on occasion flee white rule and fight against whites, good treatment could turn slaves into contented subjects, whose voluntary industry contributed to the prosperity, and therefore the security, of the empire.[84] Slaves made a direct contribution to defense by helping to build forts, and because taxes on slaves helped pay for fortification projects.[85] Freed blacks were sometimes invaluable as auxiliaries in Britain's struggles with its enemies.[86] Newspaper advertisements reminded readers that slaves regularly found and reported lost property.[87] While slaves sometimes caused problems, in fact no labour force was devoid of danger and inconvenience: advertisements in the *Gazette* reminded readers that whites committed brutal crimes in South Carolina more often than slaves did, and that white servants also ran away.[88] If South Carolina's readers were to compare their enslaved workers with the underclass in London, whose crimes and other failings were frequently discussed in the newspaper, slaves would look like a relatively safe underclass: compared to the anonymous mass of criminal poor that operated in London, runaway slaves could often be individually identified by name, distinguishing marks or character traits, and even by clothing. There was no way of enumerating, let alone identifying and controlling, the underclass of London. Finally, slaves would never become consumers of luxuries, importing foreign goods that drained the Empire of the wealth it needed for defense, as lower-class whites regularly did.

The *Gazette's* treatment of slavery demonstrated better than anything else how well the newspaper was able to portray British America as an integral part of the British Empire. Without concealing the New World's distinctiveness, the *Gazette* either subsumed these distinctions into the broader struggle being waged by Protestant Britons against Catholic enemies, or argued that America could contribute in unique ways to the Empire's economy and defense. The *Gazette* of the 1750s revealed a sophisticated portrayal of a close relationship between Great Britain and its American colonies, a portrayal where differences created a symbiotic strength, as the Empire's various parts stood firm in defense of shared values.

---

[1] Jeremy Black makes the point that eighteenth-century editors were "not simply scissors and paste men" and that selecting interesting news to build circulation made editing into a challenging, creative occupation. *The English Press in the Eighteenth Century* (Philadelphia, 1987), xiii, 35-42.

[2] Timothy H. Breen, "Ideology and Nationalism on the Eve of the American Revolution: Revisions Once More in Need of Revising," *Journal of American History*, 84.1 (June 1997), 13-39; Jack P. Greene, "Empire and Identity from the Glorious Revolution to the American Revolution," in *The Oxford History of the British Empire*, Vol. II, ed. P.J. Marshall (Oxford, 1998), 220-223.

[3] David Armitage has begun to explore the colonists' embrace of mercantilism in *The Ideological Origins of the British Empire* (Cambridge, 2000), esp. 177-184, but scholars have not yet widely accepted his arguments.

[4] *The South-Carolina Gazette*, August 17, 1752 and February 22, 1752. See also October 31, 1754 for this generic use of "English" for imperial shipping.

[5] *Ibid.*, September 12, 1754 and November 21, 1754.

[6] *Ibid.*, July 2, 1753.

[7] *Ibid.*, November 16, 1753, January 22, 1754, November 14, 1754, December 12, 1754, and November 13, 1755.

[8] *Ibid.*, November 21, 1755.

[9] *Ibid.*, May 3, 1754.

[10] *Ibid.*, January 29, 1754.

[11] *Ibid.*, January 23, 1755.

[12] *Ibid.*, January 22, 1754 and November 21, 1754.

[13] *Ibid.*, July 25, 1754.

[14] *Ibid.*, February 12, 1754.

[15] *Ibid.*, November 27, 1752.

[16] *Ibid.*, April 16, 1753.

[17] *Ibid.*, January 1, 1753, April 11, 1753, and July 11, 1754.

[18] *Ibid.*, July 2, 1753, July 30, 1753, October 15, 1753, October 22, 1753, December 3, 1753, January 22, 1754, and March 5, 1754.

[19] *Ibid.*, October 1, 1753, October 8, 1753, October 15, 1753, December 24, 1753, May 3, 1754, and August 14, 1755.
[20] *Ibid.*, January 1, 1753, July 11, 1754, and January 29, 1756.
[21] *Ibid.*, January 29, 1756.
[22] *Ibid.*, November 27, 1752, January 1, 1753, May 28, 1753, and February 6, 1755.
[23] *Ibid.*, November 16, 1753.
[24] *Ibid.*, December 17, 1753.
[25] *Ibid.*, September 3, 1753.
[26] *Ibid.*, January 30, 1755.
[27] *Ibid.*, June 20, 1754.
[28] *Ibid.*, May 3, 1754.
[29] *Ibid.*, June 4, 1753.
[30] *Ibid.*, May 18, 1753 and September 1, 1753.
[31] *Ibid.*, June 4, 1753 and April 11, 1753.
[32] *Ibid.*, October 30, 1752.
[33] *Ibid.*, May 25, 1752, April 16, 1753, and March 6, 1754.
[34] *Ibid.*, December 24, 1753.
[35] *Ibid.*, September 19, 1752.
[36] *Ibid.*, January 1, 1752.
[37] *Ibid.*, January 22, 1754.
[38] *Ibid.*, July 25, 1754.
[39] *Ibid.*, April 13, 1752.
[40] *Ibid.*, January 1, 1752 and elsewhere.
[41] *Ibid.*, February 6, 1755.
[42] *Ibid.*, January 30, 1755.
[43] *Ibid.*, February 26, 1754, March 19, 1754, and May 14, 1754.
[44] *Ibid.*, February 26, 1754.
[45] *Ibid.*, January 27, 1752, February 1, 1752, February 8, 1752, March 9, 1752, April 6, 1752, May 11, 1752, May 25, 1752, August 10, 1752, August 17, 1752, August 1, 1753, August 15, 1753, November 26, 1753, February 12, 1754, February 18, 1754, March 5, 1754, April 16, 1754, May 3, 1754, September 12, 1754, and October 10, 1754.
[46] *Ibid.*, February 22, 1752, October 3, 1752, October 30, 1752, November 20, 1752, November 27, 1752, August 15, 1753, September 10, 1753, October 1, 1753, July 4, 1754, August 15, 1754, August 29, 1754, October 31, 1754, December 12, 1754, January 16, 1755, and August 7, 1755.
[47] *Ibid.*, August 10, 1752, April 9, 1754, and August 29, 1754. South Carolinians blamed the 1739 Stono Rebellion on Spain's standing offer of liberty to slaves on South Carolina plantations who could escape to Florida. The escapees had killed and plundered whites as they attempted to reach Spanish territory and freedom.
[48] *Ibid.*, See May 28, 1753, June 12, 1753, June 25, 1753, August 6, 1753, August 20, 1753, August 15, 1754, October 3, 1754, October 17, 1754 for reports of direct French and Indian raids on South Carolina and Georgia. Reports of French activities in the Ohio were in almost every issue from June 18, 1753 through the rest of that summer, and in almost every issue after March 5, 1754.

[49] *Ibid.,* June 11, 1752.
[50] *Ibid.,* For examples of the coverage of French and Indian raids in the northern colonies, see October 10, 1752, April 23, 1753, November 16, 1753, September 19, 1754, October 17, 1754, December 26, 1754, and August 7, 1755.
[51] *Ibid.*, August 15, 1753, September 3, 1753, December 3, 1753, February 12, 1754, January 12, 1755.
[52] *Ibid.,* December 25, 1752, April 11, 1753, July 30, 1753, and December 24, 1753.
[53] *Ibid.*, June 11, 1754, July 11, 1754, July 18, 1754, and February 6, 1755.
[54] *Ibid.,* September 27, 1752.
[55] *Ibid.,* January 30, 1755.
[56] *Ibid.*, January 29, 1754.
[57] See *Ibid.*, September 17, 1753, November 16, 1753, December 17, 1753, February 20, 1755, and March 6, 1755, among other issues. For the quotation, see *ibid.,* September 24, 1753.
[58] *Ibid.*, March 12, 1754.
[59] *Ibid.,* August 27, 1753 and January 29, 1754.
[60] John Locke, *A Letter Concerning Toleration*. (Appleton-Century-Crofts, 1937), 205, 210-212. Locke argued in favor of toleration for all Protestants, but denied toleration to Catholics, because he believed they could not be trusted, even when they swore oaths of loyalty to the British government.
[61] *South-Carolina Gazette,* October 30, 1752, November 13, 1752, and March 5, 1754.
[62] *Ibid.,* September 1, 1752; August 7, 1755, August 28, 1755, September 4, 1755, September 11, 1755, October 2, 1755, October 16, 1755, October 30, 1755, November 6, 1755, and November 20, 1755.
[63] *Ibid.,* September 18, 1755.
[64] *Ibid.,* September 11, 1755 and October 9, 1755.
[65] This tradition had its roots in stories of Spanish atrocities against the natives of the New World recounted by Bartolomé de Las Casas, and it embraced the stories of the torture and burning of Protestants in England by "Bloody" Mary in the 1550s, stories that were still being excerpted in colonial publications during the eighteenth century. Colonists at this later date were also familiar with atrocity stories from the Irish uprising of 1641, which had recently been reprinted in British America under the title of *Popish Cruelty Displayed* (Boston, 1753). Similarly, the considerable number of descendents of French Huguenots in South Carolina no doubt had brought with them more believable tales of cruel persecutions suffered by their ancestors in France.
[66] *South-Carolina Gazette.,* October 30, 1752; April 23, 1753, March 26, 1754, June 4, 1754.
[67] *Ibid.,* June 11, 1754.
[68] *Ibid.,* November 13, 1755.
[69] *Ibid.,* November 20, 1755.
[70] *Ibid.,* October 2, 1754.
[71] *Ibid.,* August 7, 1755.
[72] One of the most notable examples of a historian assuming that slavery fundamentally shaped South Carolinians mental world was Henry F. May, who argued that the ideas of the Enlightenment were tinged with pessimism in South Carolina. Even when

participating in the broad intellectual culture of the Atlantic World, whites in this colony were incapable of escaping the fact that they were surrounded by the brutal and alien institution of slavery, May argued. May, *The Enlightenment in America* (Oxford, 1976), 143-149. The most significant recent work that makes slavery dominant in shaping the mental world of South Carolina's whites is Philip D. Morgan's *Slave Counterpoint: Black Culture in the Eighteenth-Century Chesapeake and Lowcountry* (Chapel Hill, 1998). Winthrop Jordan has imaginatively portrayed this nightmare world of slave owners in colonies with a black majority in *White Over Black: American Attitudes Toward the Negro, 1550-1812* (Chapel Hill, 1968), especially 110-178.

[73] *South-Carolina Gazette,* October 10, 1752 and January 8, 1753 (in Guadaloupe and Jamaica).

[74] *Ibid.,* February 26, 1754, May 14, 1754, October 17, 1754, November 7, 1754, and August 21, 1755.

[75] *Ibid.,* May 21, 1754, July 18, 1754, September 5, 1754.

[76] *Ibid.,* September 19, 1752, March 26, 1754, June 20, 1754, June 27, 1754, July 25, 1754, August 15, 1754, and August 29, 1754 (in Jamaica as well as in South Carolina).

[77] *Ibid.,* June 27, 1754 (in Albany, New York).

[78] *Ibid.,* February 26, 1754.

[79] *Ibid.,* June 27, 1754 and July 4, 1754.

[80] *Ibid.,* April 9, 1754 and August 29, 1754 (escapes or attempted escapes from South Carolina to St. Augustine), October 3, 1752 (Britons in Honduras lose slaves to Spaniards there), and January 8, 1753 (French authorities in St. Martin's refuse to return escaped slaves to Anguilla).

[81] *Ibid.,* December 5, 1754.

[82] South Carolinians had a rare experience of Catholic migration when two shiploads of forcibly dispersed Acadians arrived in Charleston in late 1755 and early 1756. The *Gazette* recorded the escape of "two Parties" of these Catholics on February 5, 1756; most were apprehended, but 30 managed to remain at large. Some robbed a plantation, terrifying the lady of the house and stealing weapons. *Ibid.,* February 5, 1756, February 12, 1756, and February 19, 1756.

[83] *Ibid.,* December 5, 1754.

[84] *Ibid.,* March 26, 1754 and May 7, 1754 (Jamaica). Governor Knowles visited towns of free blacks in the interior of Jamaica and commented that "as there seems to be a spirit of industry amongst them, I have no doubt of their continuing faithful and becoming useful to this community." These black towns had been built after the former governor, Trelawney "in person conquered the rebellious Blacks, and then treated them with so much compassion and mercy that they in gratitude built a town and called it by his name."

[85] *Ibid.,* October 10, 1752, March 26, 1754, May 14, 1754, and January 30, 1755 (examples from South Carolina and Jamaica).

[86] *Ibid.,* April 11, 1753 (in South Carolina, against "French" Indian raiders) and July 4, 1754 (against the Spaniards in Honduras).

[87] *Ibid.,* October 16, 1752, September 10, 1753, September 17, 1753, and elsewhere.

[88] *Ibid.,* Robberies and murders by whites were reported on March 12, 1754, May 21, 1754, July 4, 1754, July 11, 1754, October 3, 1754, October 17, 1754; advertisements

were regularly placed for runaway white servants, for example on September 10, 1753 and October 31, 1754.

Chapter Five

Nineteenth-Century Western Landscapes: Some British Comparisons and Contexts

Robert Grant

Images such as Albert Bierstadt's *Rocky Mountains, Lander's Peak* of 1863 [figure 1] have become iconic of the American West, exerting power over the Euro-American imagination to this day.

Figure 1: Albert Bierstadt, *Rocky Mountains, Lander's Peak*, oil on canvas, 1863 (Metropolitan Museum of Art, New York).

The promotional material the artist published to accompany the triumphant progress of such canvasses through the Eastern seaboard states and Europe, stressed the authenticity of his depictions, but he was just as clearly trading on an *idea* of the West that was arguably as important to his metropolitan audiences as the reality. Works like this, as well as a host of others by artists as diverse as Frederick Edwin Church, Asher B. Durand, Emmanuel Leutze and Thomas

Moran, elevated the American wilderness to what Barbara Novak has termed the status of a grand "natural church."[1] The images were powerfully emblematic, the overwhelming grandeur of the natural landscape signalling the nascent greatness of the American nation itself, while simultaneously suggesting a geographic *tabula rasa* on which could be written an entirely new national history. This re-absorption of wilderness to Eastern seaboard sensibilities heavily invested the American West with a set of religious resonances: as the American critic James Jackson Jarves put it in 1864, the magnificent national landscape was "the creation of the one God—his sensuous image and revelation," but it was no coincidence that these ornately gilded, carefully orchestrated and artfully staged framings of an unconflicted West reached a height of popularity during the depths of the American Civil War.[2] They seemed to suggest in their certainty and scale a transcendent, national unity founded on the "truth" of a distinctive national geography.

Such sacerdotal/triumphalist views of the American West have been seriously disputed since the 1980s.[3] And yet, while it has simultaneously become more common to situate the American West within a global context, this has too often simply reinforced the country's global power and suggested an even more aggressive exceptionalism, while eliding the subtle consonances and congruences that were produced from shared cultural outlooks which, as much as they tried to distinguish one colony or settlement from another, actually reinforced a shared Euro-American drive to global expansion. By comparing the nineteenth-century American literature and imagery of westward expansion with equivalent British material on emigration, colonisation and settlement, this chapter seeks to extend those challenges, to probe notions of American exceptionalism, to query the continuing parochialism of much American history of the West and point to the ways in which ideas about the influence of climate, changing construals of racial difference and a strain of essentially Protestant providentialism were implicated in a much larger, quintessentially Anglo-Saxon project of global expansion. Such comparisons and contexts problematise the certainties of American distinctiveness and point to shared Anglo-American experiences of nation-making.

While an American national sublime and its consonant belief in "manifest destiny" may have seen continental domination as peculiar to that nation, its language was often almost identical to that employed by advocates of British emigration, colonisation and settlement. Early nineteenth-century British enthusiasts for Australia, for example, hungered after their own distinctive form of national sublime, not least in imagining a vast interior waterway that it was believed would vie with the greatest rivers in the world.[4] If this existed, William Wentworth ruminated in 1819, "in what mighty conceptions of the future greatness and power of this colony, may we not reasonably indulge?" The vision sadly evaporated a year later when John Oxley returned from an expedition with reports of a silent, desiccated interior, and expedition after expedition as the nineteenth-century progressed confirmed the same dreadful circumstances.[5] Across the Tasman Ocean, the New Zealand landscape could, on occasion, be conceived in terms of the sublime, as in Eugene von Guérard's, *Milford Sound* [figure 2].

Figure 2: Eugene von Guérard, *Milford Sound*, oil on canvas, 1877 (Art Gallery of New South Wales).

Nevertheless, Von Guérard never settled in New Zealand. He spent only a few months sketching in the islands in the 1870s. With no art market of any size in the colony at the time, there was little point in producing anything on such a grand scale for local sale and he worked on and exhibited works like this some years later in Australia and Europe. In that respect, his work is closer to Bierstadt's: just as von Guérard returned to the metropole to stage his monumental encounters with colonial wilderness, Bierstadt returned to his New York studio and an Eastern seaboard market to stage his Western entertainments.

Of course, nineteenth-century Euro-American expansion involved the circulation of peoples, cultures, commodities and information across Europe, America and Asia, the Atlantic, Pacific and Indian Oceans, metropolis, colony and wilderness, but a word of caution is in order: it would be reductive to homogenise these views into a single, supra-personal, supra-historical outlook. The different spaces described by the writers and illustrators dealt with in this paper were the locus of disparate dramaturgies of encounter, places where different explanatory metaphors were enlisted and where different myths of encounter, accommodation and settlement were constructed.

There were no seminal voyages of discovery in New Zealand on the scale of those of John Oxley, Thomas Mitchell or Charles Sturt in Australia, for example, nor those of Zebulon Pike, Edwin James or William Keating in the United States, and the different experiences of physical scale and distance can be considered one point of departure in explaining the different responses to landscape in these different countries. New Zealand is a comparatively small country, roughly equal in size to the British Isles and relatively easily circumnavigated. Exploration of the interior could be hazardous but it did not contain the desiccated challenge that Oxley, Mitchell or Sturt faced, nor the vast expansive prairies, the immense extensions of space that permitted the exploration of the American continent to be represented as such a heroic endeavour. Finally, it lacked the climatic extremes that characterised Canada, parts of the Cape, Australia and America, and this, perhaps, made it easier to conceive of as a providential landscape, the gift of God to a diligent settler, rather than a sublime manifestation, the actual physicality of "his sensuous image."[6] It was a somewhat different Biblical metaphor derived from the Old Testament Book of Deuteronomy that best articulated this particular outlook. In Deuteronomy, Canaan is pledged to the Jewish people as a promised land after their flight from Egypt. It is described as a "land of wheat, and barely, and vines, and fig trees" and, in keeping with this image of plenty, New Zealand was more consistently depicted in the literature and art of this period in terms of agricultural, pastoral or farming landscapes.[7] On the other hand, this outlook also characterised many other colonial/settler settings: there were differences in detail but much that was common. The Cape, for example, could be conceived as an alien landscape, broken by islands of rough civilisation, sottish, torpid Boers and enslaved Khoikhoi, where violent heat was followed by equally violent thunderstorms and torrential rain. When it came to colonial/settler sites, however, the emerging prospect was similarly predicated on European and, in most cases, specifically British purchase and settlement of land. Australia was represented as a land tainted by convictism, lashed by droughts and violent bush fires, but evocations of the pastoral were characteristic of many reports of the desirable settler parts and, while Canada was often described as plagued by iron winters, dense forests and deadly miasmatic exhalations, providential opportunity characterised many of the descriptions of colonial/settler endeavour there.

Against the grandeur of Bierstadt, Church, Durand, Leutze and Moran, more quotidian images of the American West, produced in farmers' books, emigrant guides and promotional tracts such as *Prairie Scene in Illinois* [figure 3], from *The Illinois Central Railroad Company Offers for Sale Over 1,500,000 Acres Selected Farming and Woodlands*, were often very close in content, style and address to publications on British colonies.

Figure 3: Anon., *Prairie Scene in Illinois*, wood engraving, Anon., *The Illinois Central Railroad Company Offers for Sale Over 1,500,000 Acres Selected Farming and Woodlands*, Chicago, 1860 (Author's collection). [Despite the quality of this image, the author feels its inclusion is important.]

Compare, for example, *Log-house* [figure 4], from Catherine Parr Traill's *The Backwoods of Canada*. Here are the same motifs of order emerging from the inchoate, domesticity overcoming wilderness and husbandry replacing the uncultivated. The literature was also replete with comparisons across such destinations, including explicit debates about their relative merits. In 1833, for example, the British writer James Boardman recorded an American and Canadian comparing their two countries. According to the Canadian, his was nothing less than a "modern garden of Eden," with the finest houses, a diet fit for kings, an exquisite dialect, a religion that was a model for all others, crops that were the envy of her neighbours and a landscape crisscrossed by superb highways.[8]

Figure 4: Anon., *Log-house*, wood engraving, Catherine Parr Traill, *The Backwoods of Canada*, London, 1836 (Trustees of the British Museum).

Boardman did not record what the American disputant replied, but these were all tropes employed with equal spirit by American and British enthusiasts for their favoured destinations, and Americans often returned the insults. Theodore Dwight Jr described the landscape of Lower Canada as "unvarying: the inhabitants, as well as the soil, are poor, and there is nothing that deserves the name of a village."[9]

British writers could also be highly critical of the United States. William Brown contrasted the bleak, barren, sandy land near Buffalo in New York with Canada where, he pronounced, the landscape had the appearance of an English park and where signs of prosperity were found for which one would seek in vain in the United States.[10] The idea of surrendering one's "English" heritage in the United States was also a step too far for many British commentators. This was a place, according to Richard Taylor, where one lost not only one's heritage but also one's dignity, and few of the English who emigrated there reflected adequately on the differences in manners, customs and political outlook they would have to deal with, he warned.[11] Of course, for some British emigrant writers, this was the very thing that most attracted them. America had long had a reputation for accommodating the more radical emigrant, a reputation fuelled by the emigration of men like Thomas Paine and Joseph Priestley in the late eighteenth-century. This may be why, in 1809, D'Arcy Boulton marked a distinction between the political emigrant and those who sought "with greater ease, [to] maintain a rising family, and increase a small capital." He admonished the "politically motivated" emigrant to go to the United States, as "disappointed politicians" would not suit Canada and, as late as 1832, Joseph Pickering complained of demagogues trying to subvert the British constitution in Canada "by instilling democratic principles." He insisted such men should stay in the United States.[12] For Morris Birkbeck, by contrast, this was exactly what attracted him to America: what he described as a social compact based not on subjection but on combined moral and physical talents through which the good of all was promoted "in perfect accord with individual interest."[13] For other, less republican writers, the promise of a new order based on social equality was perhaps partly responsible for servants in the United States being so "saucy and difficult to please," as James Stuart put it, but social mobility (if not equality) was as much a part of the projected boon of British settler colonies as it was of America, and complaints of the absence of suitable servants were therefore common. As William Swainson observed, one of the greatest drawbacks to settling in New Zealand was the difficulty in finding and keeping good servants, but it was just as bad in the Cape, New South Wales and Tasmania, George Thompson confirmed.[14]

Conversely, American writers reflected back on the peculiar circumstances that fitted their country for what they saw as its special destiny. William Darby, on his *Tour from the City of New York, to Detroit*, played with distinctions between man and nature, wilderness and cultivation, which construed American westward expansion as both a personal endeavour and national duty. His contrast evoked, within a few paces of the cultivated farm or busy mill, "primeval silence; we could have conceived ourselves carried back to the primitive ages, when cultivation had neither disfigured or (*sic*) adorned the face of the earth."[15] British writers echoed such sentiments, although they brought with them outlooks that insisted on much grander comparisons. Boardman stood on the steps of the American Capitol and recalled a visit to its ancient equivalent with very different sensations. The old Capitol with its "solemn wrecks of grandeur" carried the mind back through the ages until it was lost in antiquity. "There every thing has been–here, on the contrary, every thing *is to be:* the reflections are all prospective."[16] Lawrence Oliphant argued the impressions of the traveller in the United States were entirely unlike those of the Old World. Instead of marvelling over decay, he must watch the progress of development. He must substitute the pleasures of anticipation for retrospection, be familiar with pecuniary speculation rather than historical association, delight in statistics not poetry, visit docks not ruins, converse about dollars not antique coin, prefer printed calico to oil paintings, and admire steam engines more than the statue of Venus. The traveller must become imbued with "go-ahead notions" to travel profitably in America, he concluded.[17] But this outlook was not unique to the United States. "'*Ennui*' is not the inhabitant of a new colony," the British writer Charles Napier retorted. In the old country, all is familiar, bound by custom.

> But in a colony all is new, all is interesting; we rise, filled with curiosity, we half shave, half wash, half dress, and then half mad, with high and joyous spirits, we jump on our horses, (our breakfast half swallowed,) and away we go.[18]

In Adelaide, South Australia, another writer enthused, the whole affair was an experiment. The newly arrived immigrant was gripped by excitement at the novelty of the situation behind which lay a determination to "make the best of it."[19]

Nevertheless, the prospective emigrant needed to be on their guard against the wildly speculative views of future prosperity and, here again, there are notable consonances between what American and British writers had to say. In 1819, the American, John Stillman Wright compiled a series of *Letters from the West: or a Caution to Emigrants*, warning of "cruel disappointment and vain regret, which so many are now enduring" in Ohio, Indiana and Illinois.[20] Some twenty years later, the British writer William Kennedy warned those thinking of emigrating to Texas to beware promoters of that particular destination, "interested eulogists," he concluded, "skilful in softening defects, or throwing them into the background, and painting whatever attractions they may possess in the colours of the rose."[21] The effect, as another British writer observed, was that many were induced to "roam about from place to place in search of an El Dorado, which is never destined to bless their eyes." At one time, the Wabash was all the rage, he noted. Then it was Illinois. Missouri became the next "grand desideratum," before it was promptly abandoned for Wisconsin, "and now, I believe, some weary with wandering about the states, have left them, and clearing the Rocky Mountains at a bound, have landed in California."[22] Stillman was no less biting in criticising his fellow Americans:

> Although the fair goddess of the terrestrial Elysium, would not unveil her beauties in Ohio or Indiana, they still hope to pay their homage at her shrine, in some favored groves: some expect to find her on the vine-fringed banks of the Arkansaw (*sic*); some amid the fragrant meads of the Red River or the Obine; while others, less sanguine, do not expect to overtake her short of St. Antony, whither, they think, she has winged her flight .[23]

Stillman, like many others, warned of "the land-jobbers, the speculators, the rich capitalists," that inflated land prices, secured monopolies on supplies, retarded the emigrant's progress or pushed them ever further in search of affordable and suitable land. Warnings against land speculation also abounded in British writings on America. William Oliver noted that many towns along the Ohio River existed in little more than name. In Kaskaskia, he espied a "splendid plan of an extensive city of the name of Downingville, with churches, public buildings, squares, &c., complete, and which, on enquiry, I found was no other than an imaginary city."[24] Still, the British colonies were no less prone to land-jobbing and the "imaginary." Louisa Meredith was amused by the carefully laid out town allotments she encountered in the wild Australian bush, "where not the semblance of a human dwelling is visible, though all arrangements seem made for a large and populous town."[25] Three days following the sale of rural lots in Auckland, New Zealand, Charles Terry reported four subdivided sections had been offered for sale as villages.

> The towns of "*Anna,*" "*Epsom,*" &c. with reserves for churches, market places, hippodromes, with crescents, terraces, and streets, named after heroes and statesmen, were the advertised, with all the technical jargon, with which colonial advertisements are characterised.

Within two weeks, ten townships had been advertised within two miles of Auckland, possessing neither roads, churches nor market places, although these were all alluringly marked out on the township plans. Under such conditions, Terry bemoaned, the country would soon be covered with roadside inns and grog houses of the lowest description, "surrounded by a few dirty hovels, inhabited by the worst of characters … and serving, as the rendezvous and abode, for all agrarian idlers and reprobates."[26]

As I've argued elsewhere, more dystopic views of colonial landscapes were just as important in re-ordering the potentially unruly, even chaotic, aspects of colonial life.[27] For would-be emigrants, the colonial frontier represented not just the possibility of a new social order, but also for the abandonment of social restraints, law and regular government, and works on both the United States and British colonies often employed a rhetoric of social and, at times racial, degeneration to make the point. Keating found the transition from an American to a French population at Fort Wayne, Indiana, unpleasant, was appalled by what he considered its "babel" and evinced the strongest disgust at the "degraded condition" of the white settlers he found there. It was Stillman's view that northerners who removed to lower Illinois "*do degenerate*:"

> It may arise from a wish to avoid giving offence to the people among whom they live, and with whom they must associate and form connexions; a love of ease may induce the abandonment of those habits of industry which the methods of farming pursued by the southern settlers, and the modes of living which prevail among them, do not require: intermarriages, too, may have their effect, or, lastly, there may exist a deteriorating principle in the very climate, which enfeebles the mental, as it *actually does*, the bodily powers.[28]

As Conevery Valencius has demonstrated, concerns about the "sickliness" or "health" of land pervaded early nineteenth-century settlers' letters, journals, newspapers and literature from the Arkansas and Missouri territories in the United States. In 1823, the American explorer Edwin James warned of bilious fevers in the upper Missouri riverine forests aggravated by the absence of physicians, want of cleanliness and the destructive habits of intemperance.[29] In 1842, the American writer Samuel Forry made a quite detailed and highly developed argument for the division of geography, climate and disease in *The Climate of the United States*. He was adamant that "the moral, intellectual, and physical capacities of man are subject to the influence of … climate."[30] In fact, prevailing Euro-American explanations of disease meant the healthfulness of different parts of the globe were discussed in remarkably consistent terms. In 1817, the British writer John Bradbury had noted that all countries "in a state of nature" were liable to cause "ague" as a result of the vast quantities of vegetable matter going into decay in Autumn, and advised precaution and prevention as the only means of dealing with them. The first settlers were liable to choose the alluvial river areas but, by doing so, inevitably sacrificed their health.[31] Right through the 1840s and 1850s, however, the phenomenon of "seasoning" continued to be frequently commented on, and its absence was as much promoted by those who favoured countries like New Zealand and Australia as it was played down by promoters of the United States. According to Godfrey Mundy, Australia had "none of the sallow and agueish faces and shaky forms the traveller meets at every step on the fertile banks of the Hooghly and the Mississippi. Even the mangrove swamps—nests of miasma elsewhere—exhale no noxious vapours in New South Wales."[32] In 1856, a correspondent from Wellington in New Zealand judged that country had a distinct advantage over the United States in having no "sickly season." This was not a country that would subject the newly arrived immigrant to the rigours of North American "seasoning," another writer enthused, the country's climate was, instead, "very favourable to the health, and development, of the human frame."[33]

All over the world, whether in America, New Zealand or the Cape, white settlers were conceived as dispelling the gloom of wilderness with aggressive self-assurance. In doing so, they swept not just what contemporary commentators saw as uncultivated wilderness before them, but also what they conceived as its uncivilised indigenous populations, which they frequently and unproblematically conflated. Edwin James, for example, collapsed indigenous animals and Native American hunters into a unified natural world lying in the wilderness between Arkansas and the Rocky Mountains.[34] William Keating mused on the evidence of the genius and perseverance of a departed Native American nation, but considered their spirit now gone. The march of civilisation was almost uniformly attended by their retreat. He listed the many tribes that had once overrun Ohio, Indiana and Illinois but who were now nearly extirpated. Only small remnants remained, and all were beginning to realise that they would be utterly exterminated if they did not turn to agricultural ways.[35] The British writer Robert Gourlay was convinced that, across the border, the interweaving of First Nation peoples with civilised society in Canada should take place as quickly as possible so that they might "be lost in that society."[36] The various failed attempts to civilise them had demonstrated to John Howison that they were a people "whose habits and characters are incapable of improvement, and not susceptible of amelioration." Their straggling numbers, wandering about the inhabited parts of Canada, had become "vicious, dissipated, and depraved," and hard drinking had impaired their senses, caused fatal combats, outrages and depravity.[37] George French Angas saw the burgeoning South Australian settlements forming the nucleus of a great empire, the "dark hunters" being driven back by the "busy hum of labour and industry."[38] Joseph Townsend also argued the Aborigines would soon altogether fade away because of an inevitable change of habits undermining their constitutions, because of vices acquired from Europeans and "abstraction" of their women by stockmen. As their fate was sealed it was only reasonable, he opined, "that what can be done to smooth their course to cold oblivion should not be omitted:" supplies of blankets would contribute to their comfort, and flour might sometimes be given.[39]

As the "savage" world opened to Europeans, the new discipline of ethnology sought to place the "civilised" world in relation to it, mapping race globally as well as increasingly in the physical features of different races, pathologising the body of the "other" by ascribing to it all that was taken as directly opposite male, European, bourgeois existence. James Cowles Prichard, for example, ascribed black coloration to "an unorganized extra-vascular substance," the *rete mucosum*, which could occur even in Europeans. He linked the darkening of white skin with pregnancy, fever, violent disruptions to normal life, even with being a beggar.[40] Black skin became the locus of a whole complex of negative associations and meanings – laziness, mental inferiority, sexual excess – that were simply givens within a field of knowledge that claimed scientific objectivity but which actually obscured the contingent workings of economic power, class and history in forming the discourse of racial difference. That discourse was dynamic, defining and redefining itself in response to ever renewed encounters with its "other" and the shifting fault lines of economic, social and cultural power in both metropolitan and colonial settings. That racial identity was represented as outside normal cultural and social boundaries was, as Drayton Nabers has argued, partly a result of growing anxiety over the fluidity of contemporary cultural identities and social norms. It was the great work of ethnology to fix that fluidity, and British ethnologists like Robert Latham engaged in immensely detailed categorisations of racial and linguistic difference. Others contended over Lamarckian laws of heritability, Cuvier's functionalism, or the uniformitarianism of Lyell, but the dominant strand, exemplified by writers like Robert Chambers, John Kenrick, Robert Knox and

Charles Smith, was concerned with the great binaries of black and white, savage and civilised, which saw racial destiny working itself out across space as well as time. As the American historian Paul Kramer has noted,

> Its rise in England was identified as only one stage in a relentless Western movement that had begun in India, had stretched into the German forests, and was playing itself out in the United States and the British Empire's settlement colonies.[41]

For some writers, as a consequence, America could be viewed as British destiny writ large. "We were Anglo-Saxon Americans," as one American commentator wrote in 1846, "[i]t was our destiny to possess and to rule this continent," and a particularly masculine Anglo-Saxonism in its westward march into the vast, supposedly uninhabited American landscape was triumphantly reflected back by Ralph Waldo Emerson's *English Traits* in 1856.[42] The British writer Lawrence Oliphant noted that the English had become accustomed to associating "the idea of an active pushing Anglo-Saxon population with the North American continent" and, as much as early nineteenth-century accounts of the American polity had offered a reproach to British social and economic conditions, the United States could also be seen as inheritor of a distinctively British liberal tradition.[43] In the 1850s, with the status of Texas and the Oregon frontier settled, links between the two countries were strengthening, not least through the economic ties of cotton production and manufacture, while continued British emigration brought greater familial connections and intellectual and literary exchange grew with the increasing facility of trans-Atlantic travel. Of course, explanations of racial difference had always circulated trans-globally, but the politics of abolitionism and the expunging of Native American title in the 1850s and 1860s made the trans-Atlantic dialogue particularly important. Indeed, by 1869, the English Member of Parliament, traveller and writer Charles Wentworth Dilke was to observe, "[t]hrough America, England is speaking to the world."[44]

The American language of "manifest destiny" was, in fact, almost identical to that employed by advocates of British expansion, and Edward Ayers has argued that American ideas about race were actually drawn largely from other colonial models, first from Spain and post-Restoration England, and then from Victorian Britain:

> in the 1850s, white Southern nationalists eagerly pored over the newspapers, journals and books of Britain and Europe, finding there raw material with which to create a vision of the South as a misunderstood place. The founders of the Confederacy saw themselves as participating in a widespread European movement, the self-determination of a people.[45]

At the same time, as Robert Young has argued, American debates about race and slavery had a material impact in Britain.[46] American ethnologists like Louis Agassiz, John Nott and George Gliddon were published in Britain, and their volumes made frequent references to their British counterparts. Trans-Atlantic exchange of a more popular type could also confirm bodies of belief about Anglo-Saxon "manifest destiny" and the inevitable disappearance of the Native American race. American artists like Samuel Hudson, John Banvard, John Rawson Smith and Henry Lewis brought panoramas of the Ohio, Missouri and Mississippi rivers to Britain and contemporary reviews reveal their brightly coloured Native American figures, wigwams and encampments were understood to effect a contrast with Euro-American settler agriculture, towns and cities that highlighted the progress of westward migration. This was also a period of great public interest in "Indian Galleries" on both sides of the Atlantic, projects that sought to capture the features of a race, as George Catlin described it in 1844, as "rapidly travelling to extinction before the destructive waves of civilization."[47]

In the United States, there was growing interest in a general census of Native Americans similar to that made in New York State between 1845 and 1846 under Henry Schoolcraft. Schoolcraft's proposal developed with congressional patronage and eventually resulted in a six volume set, *Historical and Statistical Information, Respecting the ... Indian Tribes of the United States*, produced between 1850 and 1857. Although the publication had little more than antiquarian interest (it was certainly no census), the collection, digestion and display of information on this scale was a highly visible sign of contemporary investment in the management of Native American affairs, a counterpart in its way to growing federal involvement in Native American life following the transfer, in 1849, of the United States Office of Indian Affairs from the War Department to the newly created Department of the Interior. Schoolcraft made a glib, one-sided justification for the removal of Native Americans to territory west of the Mississippi, relying on a vision of the west as a wilderness over which the existing tribes roamed, "cultivating nothing and living principally on the flesh of the buffalo," and justifying removals as necessary for the protection of Native Americans against the vitiating effects of contact with Europeans. "The Colonization Plan," as he termed the forced removals, was consummated by the evidence of improvements in "[m]orals, education, arts, and agriculture, ... and the progressive improvement in the Indian character" in the removed tribes.[48]

Schoolcraft's six volumes also coincided with the expansion of European settlement in Oregon and the south-west, a process marked by the opening of a new theatre of hostilities between Native Americans and the federal authorities. Settlement in the Oregon territory actually commenced long before Native American title had been extinguished, and signing of treaties extended well into the 1850s. At the same time, the federal government resorted to frequent force. Indian Superintendents and Agents argued conditions on the western frontier necessitated the presence of a strong military force, and the United States army took an active part in punitive raids, as well as larger confrontations with Native Americans such as the Yakima War of 1855, the Battle of Seattle in 1856 and war against the Spokanes in 1858. Reports of these conflicts amplified the longstanding belief in American westward expansion as a contest between races, "of civilization against savageism," as Schoolcraft put it, "and, as in all conflicts of a superior with an inferior condition, the latter must in the end succumb. The higher type must wield the sceptre."[49] The western horizon was the new horizon of western civilisation, but it was a horizon of European violence and the sunset of many Native American lifeways.

As a number of writers have argued, by focussing on providentialism, racial predestination and ideas that the American continent was the "natural" inheritance of the American nation, "manifest destiny" rationalised a deliberate federal policy of Native American removals and the use of ruthless military force in the conquest of the American West.[50] The American sense of a divinely sanctioned conquest was rooted in the first, brutal century of English incursion into the American continent but, while American histories generally trace the roots of "manifest destiny" to those first contacts, as a specifically British experience, it also laid the foundation for the slaying of Aborigines in Australia, for the conquest of Xhosa, San and Khoikhoi in the Cape, and for the destruction of Maori lifeways in New Zealand. Far from being an innate drive in the Euro-American population, it was a self-conscious creation of a group of political propagandists who drew on and amplified a core of longstanding tropes, global in their extent and applied to the most varied landscapes from America to the Cape, Canada to New Zealand, but always forming a *leitmotif* that glossed the violence of European encroachments on indigenous populations, flora, fauna and landscapes.

---

[1] Barbara Novak, *Nature and Culture* (Oxford, 1981), 151.

[2] James Jackson Jarves, *The Art Idea* (New York, 1864), 86.

[3] See, for example, William Cronon, George Miles & Jay Gitlin, eds. *Under an Open Sky* (New York, 1992); Patricia Nelson Limerick, Clyde Milner II & Charles Rankin, eds. *Trails* (Lawrence, 1991;) Richard White, *"It's Your Misfortune and None of My Own"* (Norman, 1991).

[4] See, for example, George Grey, *Two Expeditions of Discovery in North-West and Western Australia*, 2 vols. (London, 1841), I 289-90.

[5] William Wentworth, *Statistical, Historical, and Political Description of ... New South Wales* (London, 1819), 77.

[6] Novak, *Nature and Culture*, 151

[7] See for example, Arthur Thomson, *The Story of New Zealand* (London, 1859), 316; Charles Hursthouse, *New Zealand, the Britain of the South* (London, 1861), xi; Henry Petre, *Account of the Settlements of the New Zealand Company*, 5th edn. (London, 1842), 30; Edward Fitton, *New Zealand* (London, 1856), 203.
[8] James Boardman, *America* (London, 1833), 338-9.
[9] Theodore Dwight Jr., *The Northern Traveler* (New York, 1830), 196.
[10] William Brown, *America* (Leeds, 1849). 77-9.
[11] Richard Taylor, *Te Ika a Maui* (London, 1855), 458-459.
[12] D'Arcy Boulton, *Sketch of His Majesty's Province of Upper Canada* (London, 1805), 4; Joseph Pickering, *Inquiries of an Emigrant* (London, 1832), ix.
[13] Morris Birkbeck, *Journey in America* (London, 1818), 109-10.
[14] James Stuart, *Three Years in North America*, 2 vols. (Edinburgh, 1833), II, 221; William Swainson, *New Zealand and its Colonization* (London, 1859), 227; George Thompson, *Travels and Adventures in Southern Africa*, 2 vols. (London, 1827), II, 124.
[15] William Darby, *Tour from ... New York, to Detroit* (New York, 1819), 31, 58.
[16] James Boardman,. *America* (London, 1833), 234 original emphasis.
[17] Lawrence Oliphant, *Minnesota and the Far West* (Edinburgh & London, 1855), 1-2.
[18] Charles Napier, *Colonization* (London, 1835), 78.
[19] E. Lloyd, *A Visit to the Antipodes* (London, 1846), 88.
[20] John Stillman Wright, *Letters from the West* (Salem, 1819), ix.
[21] William Kennedy, *Texas: The Rise, Progress and Prospects*, 2 vols (London, 1841), I, 80.
[22] William Oliver, *Eight Months in Illinois* (Newcastle upon Tyne, 1843), 139.
[23] Wright, *Letters from the West*, 41
[24] Oliver, *Eight Months in Illinois,* 13, 21.
[25] Louisa Meredith, *Notes and Sketches of New South Wales* (London, 1844), 53
[26] Terry, Charles. *New Zealand* (London, 1842), 155 & 161-2.
[27] Robert Grant, "'Delusive Dreams of Fruitfulness and Plenty': Some Aspects of British Frontier Semiology c.1800-1850'", *Deterritorialisation*, Mark Dorrian & Gillian Rose, eds. (London, 2003).
[28] Wright, *Letters from the West,* 34-5 original emphases.
[29] Edwin James, *Account of an Expedition from Pittsburgh to the Rocky Mountains*, 3 vols. (London, 1823), I 110, II 262-3.
[30] Samuel Forry, *The Climate of the United States* (New York, 1842), 21.
[31] James Bradbury, *Travels in ... America* (London, 1817), 328.
[32] Godfrey Mundy, *Our Antipodes*, 3 vols. (London, 1852), I, 256
[33] Fitton, Edward. *New Zealand* (London, 1856), 345; Ward, John. *Information Relative to New Zealand*, 2nd edn. (London, 1840), 18, See also Robert Grant, "New Zealand 'naturally': Ernst Dieffenbach, environmental determinism and the mid nineteenth-century British colonisation of New Zealand," *New Zealand Journal of History*, vol. 37, no. 1 (April 2003).
[34] James, *Account of an Expedition,* II, 161

[35] William H. Keating, *Narrative of an Expedition to the source of St. Peter's River.* (Philadelphia, 1824), (II 240, I 39, 83, 116, 230 & 422).
[36] Robert Gourlay, *Statistical Account of Upper Canada*, 2 vols. (London, 1822), II, 392.
[37] John Howison, *Sketches of Upper Canada* (London, 1821), 148, 151.
[38] George French Angas, *South Australia* (London, 1846), preface.
[39] Joseph Townsend, *Rambles and Observations in New South Wales* (London, 1849), 107.
[40] James Prichard, *Researches into the Physical History of Mankind*, 4th edn. 5 vols. (London, 1837-1845). I, 234-6
[41] Paul Kramer, "Empires, Exceptions, and Anglo-Saxons," *Journal of American History*, vol. 88, no. 4 (March 2002), 1322. See also Reginald Horsman, "Origins of Racial Anglo-Saxonism in Great Britain before 1850," *Journal of the History of Ideas*, no. 37 (1976) pp. 387-410; Hugh McDougall, *Racial Myth in English History* (Hanover, 1982).
[42] Ayres, William. ed., *Picturing History* (New York, 1993), 207.
[43] Oliphant, Lawrence. *Minnesota and the Far West* (Edinburgh & London, 1855), 21.
[44] Charles Dilke, *Greater Britain* (London, 1869), viii.
[45] Cited in Ann Stoler, "Tense and Tender Ties," *Journal of American History*, vol. 88, no. 3 (December 2001), 849.
[46] Robert Young, *Colonial Desire* (London, 1995), 124.
[47] George Catlin, *Indian Portfolio* (London, 1844), 5.
[48] Henry Schoolcraft, *Historical and Statistical Information, Respecting the ... Indian Tribes of the United States* 6 vols (Philadelphia, 1851-1857), VI 428-506 & 512.
[49] Ibid., VI, 28
[50] On Native American responses to both the doctrine and realities of 'manifest destiny', see Deborah Madsen, *American Exceptionalism* (Edinburgh, 1998), 41-69.

CHAPTER SIX

DICKENS'S AMPUTATED AMERICA

MAX STITES

In 1842 Charles Dickens wrote from the United States: "This is not the Republic I came to see. This is not the Republic of my imagination."[1] Not only does this statement make clear the novelist's construction of an imaginary America prior to visiting the country, it also begs the question of precisely how Dickens envisaged the Republic that, to his distress, did not exist on the western shores of the Atlantic. Scholars have often sought to study Dickens's imaginary republic through some combination of the novelist's letters, his travelogue *American Notes*, and *Martin Chuzzlewit*. Harry Stone, for example, has convincingly argued that these three sources "form an extraordinary trilogy of materials made to order for the study of the relationship between fact and fiction."[2] As this observation suggests, though such efforts do provide valuable insight into Dickens's expectations of and experience with the United States, they also tend to focus more on comparisons between his perceptions of the relationship between American ideals and realities rather than precisely how Dickens imagined the country. This trio of letters, travelogue and novel may illustrate Dickens's frustrated reaction to what he viewed as the failure of the United States to meet its potential, but it does not necessarily provide a complete picture of how the Englishman constructed the country in his mind.

Considering the character of Joe Willet in *Barnaby Rudge* in addition to these three sources helps to uncover a more nuanced view of how Dickens imagined the Republic, revealing his conception of America as a locus for a range of conflicting forces and potentials. Joe Willet and the arm he loses in America serve as the central image of this argument, embodying the idea of the former British colonies as severed from a larger political entity. With this image of an "amputated America" in mind it becomes possible to view the events of Dickens's letters from America, *American Notes* and *Martin Chuzzlewit* as more than the expression of one author's angry disillusion with what he saw as the failure of the United States to meet the ideals he had associated with it. Willet's symbolic amputation allows such comparisons to be set aside allowing a view of

America imagined as the encapsulation of a range of ideals and ideas of, among other things, opportunity, freedom and violence to begin to emerge.

Given that in *Barnaby Rudge* Joe's American adventure takes place outside the directly reported events of the narrative it may appear incongruous to invest his time there, and the arm he loses with such significance. Such incongruity disappears in the face of Joe's reasons for journeying to the USA, his activity there and the historical context of his journey. When Joe Willet runs away, he does so not merely out of a boyish desire for excitement and glory, but because he can no longer silently endure the tyranny of John Willet and his cronies. This is not to suggest, however, that dreams of magnificence play no part in the young man's actions. Joe himself reveals his own romanticised view of the military when, in response to an innkeeper's suggestion that soldiers who are shot "are the best off," he replies to the man, "but you don't care for glory."[3] With these words, the young Willet makes clear that whatever else might motivate him he possesses no immunity from stereotypically adolescent dreams of military life. By the time he makes this statement, however, Joe has already fled his home because of his ill treatment there, suggesting that whatever glorious dreams he might possess, those dreams do not serve as the primary impulse behind his actions. Instead, Joe only gives implicit voice to this romanticised view of military life after being forced to pursue a desperate course of action in order to free himself from constant subjugation, abuse and ridicule.

When the weight of the degradation suffered by the young man finally becomes too much for him to bear he attacks his father's crowing friend, Mr Cobb. Readers are told that, "Crowding into one moment the vexation and the wrath of years, Joe started up, overturned the table, fell upon his long enemy".[4] On one hand, this violence results from Joe's long-simmering anger at his subjugation finally reaching a full boil. On the other hand, the suggestion that "the vexation and wrath of years" explodes in a moment of rebelliousness lends a sense of justified inevitability to the outburst which, viewed alongside Joe's eventual flight to America, connects his personal travails with ideas associated with America and its revolt against Britain. Like Joe, the thirteen American colonies could be seen as rebelling against a catalogue of abuses endured over time at the hands of an archetypal patriarch in the form of a King. The colonial rebellion could also be construed as a sudden eruption of anger and self-assertion after, as in Joe's case, the silent endurance of patriarchal abuse becomes impossible. In both cases, some sort of father figure prompts rebellion through the misuse of paternal power and authority.

In fact, the image of Britain as a parent to the United States is one deployed elsewhere by Dickens when an American asks Martin Chuzzlewit in reference to Britain: "how's the unnat'ral old parent by this time?"[5] Not only does this later positing of a parent-child relationship between Britain and the former North American colonies make clear the novelist's willingness to imagine a definite kinship between the United States and Britain, it also offers an important clue as to the correct interpretation of Joe's flight from home and his later loss of limb. Joe flees his home because his father mistreats him thereby failing in his duties to his son. This same sense of fraternal or sanguinary obligation resonates in the later reference to Britain as the parent of the United States. John Willet, like Britain, emerges as somehow unnatural, while his failings as a father do more than simply justify his son's flight. They also add symbolic weight to Joe's American sojourn by allowing this individual father-son relationship to echo the larger, political relationship between Britain and the United States through the metaphor of parenthood. As a result, Joe's journey to America not only demonstrates Dickens's willingness to imagine the United States as a place of escape from injustice and curtailed liberty, but also suggests the possibility that the novelist may have imagined the American war of Independence in the glorious terms of justice and liberty suggested by American rhetoric. Like Joe, the North American colonies could be viewed as fleeing from and revolting against unnatural and harmful treatment at the hands of the patriarchal powers and institutions that should have cared for them. Once again, such an impulse does not preclude more glorious dreams of magnificence, but it does lend an air of justifiability to such rebellion. Through this depiction of failed parental responsibility and Joe's flight to America then, Dickens might be assumed to have simply accepted the more positive renderings of the American Revolution as an unqualified benefit to America and even the world.

As the rest of Joe's story makes clear, this is not the case. Before turning to that story, it should be noted that Joe is only one example of a Dickensian character whose story suggests a view of the United States as a place of escape from tyrannical fathers and institutions. Though the novelist did not always make use of the parental metaphor elsewhere, examples in his other works indicate his continuing fascination with the United States as a kind of safety valve or escape mechanism for the harassed and downtrodden. Tony Weller, for example, famously suggests that Mr Pickwick could avoid the consequences of a breach-of–promise suit by being smuggled out of Fleet Prison in a piano, and then fleeing to America. Mr Weller further extols this plan by suggesting: "and then let him [Mr Pickwick] come back and write a book about the 'Merrikins as'll pay all his expenses and more, if he blows 'em up enough."[6] In this case, the United States serves not only as a place of safe escape from the persecutions of the law, but as a source of potential material wealth as well. The fact that attainment of wealth in the New World lies, according to Weller, in the process of writing about America as a country and a people should not be overlooked. On one hand, the suggestion that such an exercise could pay Pickwick's

American expenses if properly pitched to assuage American vanity amounts to a somewhat cutting judgement of the American character as egotistical and in need of stroking. More importantly, this comment suggests, on the other hand, that even at a stage when Dickens had no concrete plans to visit the young Republic he viewed it as a potentially fruitful subject that held particular promise for him as an author. By appealing to Mr Pickwick as someone capable of writing a commercially successful book about the United States based on first-hand experience, Tony Weller betrays the possibility that Dickens viewed himself as capable of special insight into the country as a writer.

As the novelist's correspondence makes clear, this confidence in his own powers of insight into the young Republic made a stronger showing as he prepared for his 1842 visit to the country. In October of 1841 he wrote that:

> My notion is that in going to a New World one must for a time utterly forget, and put out of sight the Old one and bring none of its customs or observances into the comparison—Or if you do compare remember how much brutality you may see (if you choose) in the common streets and public places of London.[7]

Such a statement clearly amounts to an admirable recognition of the necessity of open-mindedness while travelling as well as constituting an expression of a degree of prescient cultural relativism. This statement not only reveals Dickens's desire to view the United States fairly, it also recognises the way in which familiarity with the vice and corruption of a traveller's native land breeds a blindness to their existence which has a deleterious affect on any attempt to compare customs and cultures. At the same time, the clearly didactic tone of the passage allows Dickens's sense of his own authority on the matter to emerge. The implicit corollary to the wisdom offered in this letter is Dickens's supreme confidence that he, more than other travel writers, is precisely the individual able to view the United States on its own terms. Furthermore, Dickens undoubtedly has complete faith in his unobstructed vision of his own country and culture when considered alongside that of the US. This passage, in other words, puts the implications of America as a fertile land for the mind of an author into explicit terms with specific applicability to the creator of Tony Weller and Samuel Pickwick.

While this consideration of the novelist's assumption of his own unique power of insight into America may appear somewhat digressive, it is significant in that it makes clear that both at the time of the composition of *The Pickwick Papers* and beyond, Dickens did not remain restricted to a view of the United States as nothing more than a place of escape from fathers and institutions. Instead, the literary element of Tony Weller's plan, combined with the letter to Bell, makes clear that the United States exerted a continuous pull on the mind of the novelist as a subject worthy of his imaginative and literary attentions. Unlike his author, however, Mr Pickwick never makes the journey to America, nor does he produce the tome suggested by the elder Weller, indicating that even at this early stage, Dickens's imaginary republic was nothing as simple as a universal safe haven from persecution and prosecution in Britain. Instead, Samuel Pickwick remains in Britain to face the consequences of unjust legal proceedings. By doing so he suggests a view of the United States that recognizes the country as a potential means of escape and land of freedom because of its political and institutional severance from Britain, but not as a panacea. Viewed alongside the material wealth America might offer literary practitioners, this rejection of the image of America as a universally available safe haven makes clear that Dickens, despite any potentially over-weaning confidence, was capable of imagining America in terms of shadings and gradations rather than absolutes. Rejection of America as a universally accessible safety valve or place of escape has great relevance to a more complete understanding of how Dickens imagined the country that should not be overlooked.

This refutation becomes increasingly explicit through both Joe Willet, and the younger Martin Chuzzlewit. Like Joe, Martin leaves Britain in search of better fortune having struggled against a series of injustices stemming from his troubled relationship with a patriarchal figure. In Martin's case his eponymous grandfather serves as the patriarchal impulse to his flight rather than the man who sired him, further reinforcing the argument that Dickens exploits the familial duty implicit in such relationships on a variety of levels. Martin's expectations are also more explicitly stated than those of Joe, partly because while Willet enlists merely with the aim of going abroad, Martin specifically chooses the US as his destination. Additionally, Dickens's decision to keep his readers in suspense about Joe's fate prevented him from referring to the young man's destination after enlistment in terms more specific than "out of London" and "a country where it's all sunshine and plunder."[8] Such a description does make clear that wherever Joe winds up he can expect, or so he is led to believe, greater comfort and potential material success than he does in Britain. It does not, however, suggest the possibility of anything other than anonymous advancement, while this reference's vagueness as to geographical location suggests that the principal attraction of this sunlit land lies more in the fact of its distance from Britain than anything else. Joe's America, in other words, remains a vaguely defined place of escape that appears to function more through a freedom from the strictures of his native country rather than any possibility of providing a fertile environment for the development of his particular talents.

In contrast to this, once Martin has fixated on America as his destination he makes clear the precise nature of the hope and opportunity he sees the country as offering to him. As Martin tries to find means of passage to the New World readers are told that: "It is an illustration of a very common tendency in the mind of man, that all this time he never once doubted, one may almost say the certainty of his doing great things in the New World, if he could only get there."[9] Martin sees the United States as a setting for his own pursuit of success and recognition not merely a place of escape. His expectations of greatness reveal that the hopeful future he locates for himself in the United States lies in more than a freedom from his past, though, as with Joe, this remains a strong element of his optimism. Accompanying this recognition of freedom from the limitations of Britain comes a strong sense that somehow America will not only foster, but welcome whatever abilities Martin may possess. Gone is the sense of flight into a liberating anonymity that pervades Joe's imaginary haven. Such anonymity is instead replaced by Martin's egotistical confidence that in America he will achieve success and recognition quickly; in his mind the US is just as eager for him as he is for it. Martin provides further emphasis of this fact when he tells his betrothed, Mary: "I am going to America, with great prospects of doing well, and of returning home myself very soon;"[10] making doubly clear just how rosy he believes his future to be. Young Joe Willet never finds himself capable of uttering such sentiments to himself or anyone else.

While Willet's and Martin's expectations of what awaits them when they leave Britain do differ in this important aspect, it should be noted that they also indicate Dickens's awareness of a popular, if general mode of imagining the United States. The positing of Martin's own expectations as an example of a widely displayed tendency points to this, through the acknowledgement of the wider influence of the forces at work on the young man. Partly, this amounts to a clever way of saying that Martin, like many others, has fallen prey to a belief that the grass is, in fact, greener on the other side of the Atlantic Ocean. Combining this fact with the realisation that Martin, like Joe Willet, seeks a place of freedom from the restrictions imposed upon him by a patriarchal force in Britain points to a significant conclusion. Though the two characters may have somewhat divergent expectations of what awaits them in their search for escape and freedom, they both make clear that Dickens consciously deployed an archetypal pattern for imagining the United States in each of their cases: that of a new start in the west. The pervasiveness of this pattern becomes clear when it is recognised that as Fossum and Roth among others have argued: "the motif of new beginnings is tied to the image of America as the New World, a potential New Eden in the West, as well as common American attitudes toward history, opportunity, success and failure."[11] Not only do both of these characters and their stories include a recognition and pursuit of some form of new beginnings in America, they also conform to these critics' general definition of attitudes they consider specifically American. As discussed above, Joe and Martin both seek a release from the restrictions imposed upon them by patriarchal power.

They both seek, in other words, an escape from some element of their respective pasts.

Their shared belief that the past can be escaped by leaving Britain, and the fact that both leave that country for America therefore provides a significant illustration of, and counterpoint, to the idea that "From their inception, American self-images reflected the idea that the past did not bind one irrevocably."[12] Martin and Joe clearly share this idea despite the fact the neither of them are American. Viewed alongside their mutual illustration of the motif of new beginnings, their respective searches for a life free from the strictures of the past that carry them to the United States make clear that Dickens not only possessed an awareness of America imagined as a place of new beginnings, but that he recognised and gave expression to an British susceptibility to this motif that Fossum and Roth do not appear to consider at any depth. This pair may recognise that the American Dream of new beginnings has European roots but it is Dickens who explores how the alluring image of America as a place to make a fresh start exerted an imaginative pull on young Englishmen like Joe and Martin.[13] They may not share precisely the same expectations of the United States, but they both allow insight into an British imagining of the country as a place of freedom from the past and potential success in the future because of its severance from Britain. At the same time, Dickens is careful not to leave such illustrations of an British belief in an America of new beginnings unqualified. Despite differences in their views, both young men return home from the United States having experienced hardship: Joe has lost an arm; Martin his money, pride, and, very nearly, his life, while the fact that Martin's hardships occur in a place called Eden, makes doubly clear that Dickens made conscious use of the motif of new beginnings. These broad facts alone provide enough evidence to suggest that in imagining America Dickens did not subscribe to any concrete or absolute views of United States as an unqualified good or a boundless evil, a place of unhampered beginnings or malign endings.

From Mr Pickwick and Joe Willet, to Martin Chuzzlewit the general narrative circumstances of the novels they appear in imply a view of America that is shaded by contradictions. While some critics have, in reference to *Martin Chuzzlewit*, rightly argued that by the time he wrote the novel, "Dickens [was] clearly interested in generalizing about American failings,"[14] to accept such a statement as definitive of the novelist's imaginative engagement with the United States is, in certain instances, to risk missing a broader range of ideas that Dickens associated with the country. A passage from *Martin Chuzzlewit* helps to illustrate that even in the midst of his generalizations about American failings, Dickens retained some concept of the country as a locus for a range of possibly conflicting ideas. As he and Martin depart from the New World, Mark Tapley describes how he would draw the American Eagle with the words:

> like a Bat, for its short-sightedness; like a Bantam, for its bragging; like a Magpie, for its honesty; like a Peacock for its vanity, like an Ostrich, for putting its head in the mud and thinking nobody sees—.[15]

This critical rendering of an American icon is both cut short and countered by Martin's response "And like a Phoenix, for its power of springing from the ashes of its faults and vices, and soaring up anew into the sky!"[16] making clear that for character and novelist alike, even intense disillusion with perceived American shortcomings do not suffice to fix narrowly the country's value as an actual place, a symbol or an imaginative construct. Martin after all, has had an intensely difficult time in the United States. His early confidence has been shattered through the loss of his money and, very nearly, his life. If any of Dickens's creations might be expected to have nothing to offer but criticisms of the United States, that character would be Martin. As a result of this, Martin's voicing of the simile equating the American eagle with a phoenix underscores a concept of America that has enough flexibility and breadth to encompass optimism and pessimism. Martin more than anyone else in Dickens's oeuvre has a first-hand knowledge of the smouldering remains of the nation's shortcomings. Though he does not dismiss these faults, he does allow for some sort of regenerative possibility that, through the image of the phoenix, provides subtle, even poetic, echoes of American new beginnings through an escape from the detritus of the past.

Moreover, it should not be overlooked that through this passage offering a new interpretation of a significant American icon, the novelist once again signals his awareness and deft handling of archetypal imaginative renderings of the United States. Though the eagle provides a more tangible and concrete image of the country than the motif of new beginnings, it nonetheless resonates with imaginative and symbolic power. By having Mark Tapley engineer a new American eagle that is a hybrid of a variety of other birds, the novelist not only signals his awareness of American symbolism but, through its subversion, offers his readers a glimpse of how such symbols function in his own imaginative rendering of the country. As with the new beginnings motif, Dickens once again illuminates an intersection between modes of perception more commonly considered American and British imaginative capacities. In doing so, Dickens may paint a critical picture of the young Republic, but he never dismisses the possibility of some redemptive force being at work there.

Of course, Martin and Mark's attempt to describe the American eagle does not serve as an explanation of how Dickens imagined America beyond cataloguing certain national tendencies in addition to the suggestion of ambiguity and ambivalence. Nor does the recognition of characters such as Samuel Pickwick, Joe Willet and Martin Chuzzlewit as representatives of a qualified view of America as a potential place of escape do much to clarify the author's views fully. Instead, in order to reach a greater understanding of Dickens's imaginary republic, we must return to Joe Willet and his amputated arm. When Joe reappears in *Barnaby Rudge* after a lengthy absence in the service of the British military, some time elapses before his father and readers learn how he lost his limb. When John Willet finally expresses his understanding that Joe's arm has "been took off!" Joe explains that his limb was lost "At the defence of the Savannah . . . In America, where the war is."[17] These brief comments, more than any others, indicate the precise nature of how Dickens imagined the United States both at the time he was writing *Barnaby Rudge* and beyond. Joe's loss of his arm in a war that eventually resulted in American independence from Britain makes clear his status as a synedochal representation of Britain's relationship with its former North American colonies. What Joe Willet's missing arm and the circumstances of how he lost it illustrate is that for the novelist the United States was a political appendage severed from a larger British body, an "amputated America." Joe represents the nation and institutions of his native Britain, while his missing arm (presumably left behind someplace in America) stands for the former colonies that became the United States through a violent act of severance. Such imagery not only makes clear Dickens's view of the young Republic as an "amputated America," it also reveals something of the emotional tenor of the construct as a way of imagining the United States. Amputation implies a certain amount of trauma through the forcible division of two parts of a larger body. This suggests that, for Dickens, America's severance from its parent nation was, at some level, inseparable from a disruption of proper order and organisation despite all its potential benefits. While this may be the case, it is important to keep in mind that such an image does not apportion blame for the split to either Britain or the United States. A patient who undergoes amputation, after all, may bear complete responsibility for the circumstance resulting in the operation, or none at all.

Rod Mengham has written of the American War of Independence that: "For the British forces, the American war was another conflict over property, the military campaign an attempt to forestall the loss of British possessions," thereby providing the necessary historical context for viewing the younger Willet and his missing limb as an imaginative rendering of the relationship between Britain and the United States.[18] By serving in the British armed forces at the "defence of the Savannah" and losing a limb Joe not only participates in but also prefigures the more general "loss of British possessions" that the war became in human, corporeal terms. The image of amputation also helps to suggest that, at some level, the war was not simply a conflict over property, but instead intimately tied to the characteristics rendering people human. Though the image of Joe's lost arm undoubtedly includes the loss of property that Mengham refers to, it also suggest a traumatic emotional experience above and beyond the loss of mere possessions that sounds a subtle note of doubt about the separation of the United States from Britain. If, as is tempting, the prefiguring role played by the younger Willet is taken to indicate a view of American independence as some form of tragic loss, an unhappily amputated America as it were, it may appear an incongruous role for this character to play. Joe has run away from home in a fit of justified rebellion against unfair treatment at the hands of his father and his father's friends. On their own these circumstances not only appear to justify the son's actions, they might also be seen as opposing a view that questions America's separation from Britain through the image of amputation.

This possibility of opposition, or incongruity, is brought into greater relief by the fact that Joe fights for the British to prevent colonial independence. Given that the narrative of *Barnaby Rudge* never subverts Joe's decision to flee his father's abusive and restrictive wielding of patriarchal power, readers might reasonably see Joe as a more appropriate ally of the Americans in their attempt to escape the impositions of another such authority. While Joe's reasons for his flight from home do suggest rebellion against despotic authority is excusable, such rebellion, according to the novel, must remain within certain limits. Joe eventually returns home and re-establishes a relationship with his father on a more equitable footing thereby illustrating Mengham's assertion that "the vital principle upheld in the elaboration of family histories in the novel is that of succession, of the effective transfer of responsibility from one generation to the next."[19] Not only does Dickens use the image of amputation to suggest a more significant bond between the United States and Britain than mere political affiliation, but, as seen above, he was also willing to consider the two countries as parent and child. Amputation and the metaphor of blood relation suggest that both order and succession have veered from their proper channels, while Joe's own circumstances and their happy resolution after his return to Britain implies the necessity of upsetting that same order and succession at least temporarily. At this stage, it appears that Dickens offers little indication as to how to reconcile such opposing views.

By turning to John Willet's reaction to his son's missing limb it becomes possible to glimpse how the Dickensian construct of an "amputated America" can contain apparently opposing impulses such as justified rebellion against patriarchal tyranny, as well as an impetus to maintain order and lawful possession. During a dinner at the Black Lion, the very place where Joe enlisted, John Willet repeats his son's explanation of how he lost his arm saying, "It was took off in the defence of the Salwanners in America where the war is."[20] On its own, this repetition makes clear the Willet patriarch's inability to reconcile himself to his son's amputation through a specific illustration of the earlier general assertion by the narrator that: "The subject that worked in Mr Willet's mind . . . was no other than his son's bodily disfigurement, which he had never yet got himself thoroughly to believe of or comprehend."[21] When this general statement and John's repetition of his son's words come under consideration alongside the revelation that:

> Continuing to repeat these words to himself in a low tone of voice (the same information had been conveyed to him in the same terms, at least fifty times before), Mr Willet arose from the table; walked round to Joe, felt his empty sleeve all the way up, from the cuff, to where the stump of his arm remained [22]

a more significant structure emerges. John's parroting of his son's words clearly illuminates how strange the loss of an arm has rendered the son in the eyes of the father. It also firmly aligns that strangeness and its accompanying incomprehensibility with the physical trait that cements Joe in his role as the representative of an "amputated America."

In feeling the "empty sleeve all the way up," John Willet betrays his failure to fully recognise his altered son by resorting to the sense of touch to aid his sight and intellect. Some part of John's sluggish intelligence clearly comprehends that Joe has lost an arm as he repeats the explanation he has been given. The fact that this repetition accompanies a more fundamental physical examination, meanwhile, suggests that such abstract mental comprehension does not necessarily entail John's full emotional or instinctual recognition of his altered family member. Through the symbolic significance invested in Joe and his "stump" this failure represents a similar difficulty in recognising and comprehending a United States that has been severed from the larger political body of Great Britain. Whether in the case of an actual father looking at his son, or that of a more metaphorical patriarch such as a British subject, viewing that renegade offspring, the United States, the very act of severance, of amputation, and all the emotional connotations such an act implies, makes any attempt at understanding uncertain and unstable. Severance from Britain clearly renders the United States a partially unknown and unknowable quantity for Dickens. The very act of amputation makes America something alien than can never be fully reintegrated into the larger political body or the imagination just as Joe's arm cannot be reattached to his body. It is this uncertainty that finally allows the nuances and ambivalences of Dickens's amputated America both before and after his 1842 visit to the country to exist side by side, rather than in binary opposition.

In the case of both the United States and Joe Willet such uncertainty and ambivalence stems, in part, from confusion as to where the ultimate responsibility for the act of severance lies. John Willet is described as:

> having certain vague misgivings within him, that he [Joe] was ready on the shortest notice, and on receipt of the slightest offence, to fell the Black Lion to

> the floor of his own parlour, and immediately withdraw to China or some other remote and unknown region, there to dwell forever more; or at least until he had got rid of his remaining arm and both legs, and perhaps an eye or so into the bargain.[23]

This passage includes a certain grotesquely comical element through John's apprehension that his son might flee Britain once again with some dark determination to leave as many of his body parts as possible on foreign soil. Despite such humour, John Willet's fear that Joe may make a second flight is neither dismissed nor confirmed. Instead, the very fact that the older Willet worries that he may be capable of giving his son an affront that will precipitate further drastic action amounts to a recognition that, at some level, he might bear some responsibility not just for Joe's original flight, but also for the loss of limb it resulted in. This possibility is countered both by the darkly humorous image of Joe divesting himself of his remaining limbs in a fit of pique as well as the suggestion that the younger Willet may be on a sharp lookout for any possible offence to which he can respond. This passage highlights, in other words, the fact that Joe's flight from his father may have been justifiable and that some blame for what ensued can be attributed to John Willet. At the same time, it tempers that justifiability and guilt with Joe's potential over-sensitivity and personal responsibility for the loss of his arm.

Given that this confusion centres on the missing limb that allows Joe to serve as an embodiment of the relationship between Britain and the United States, it is apparent that the novelist is doing more than commenting upon Joe, his missing limb and John Willet. Like John, Britain may have held a measure of responsibility for the revolt of the colonies that became the United States, while those same colonies may have suffered from thin skins and quick tempers. The indeterminacy of where responsibility lies for the separation of limb from body, colonies from country amplifies the uncanny ambivalence that the simple fact of such severance already carries with it. Rather than demonstrating a concern with the proper apportionment of guilt, Dickens makes use of Joe's amputation and the emotional and mental responses it provokes in his father to signal his understanding of the United States as a former part of Britain and as a free-standing entity. Significantly, the confused anxiety endured by John Willet plays the double role of underscoring the alien nature of amputation as well as signalling the possible location of blame for the occurrence without ever resolving the issue. Thus Dickens demonstrates a greater concern with exploring the finer nuances of Joe's amputation and the political relationship it symbolises rather than trying to dispel the appearance of contradiction in favour of a final judgement. Neither Joe nor the United States emerge as the definitively guilty or innocently wounded party in this symbolic exploration of the outcome of their revolt against a patriarchal power.

Instead America comes to exist as a range of potentials and traits best illustrated by Mark and Martin's exchange about the American Eagle as it summarises a variety of negative American characteristics only to end with the positive, if somewhat ambiguous, image of the phoenix rising from the ashes of its own destruction. It is worth noting, by way of a slight digression, that while the loss of an arm may not constitute destruction, Joe's story does contain a subtle suggestion of loss and rebirth. After all, it is only after Joe's return to Britain "as poor in purse as [he] went, and crippled for life besides,"[24] that he wins the hand of Dolly Varden to live happily ever after in a rather typical Dickensian mode of domestic contentment. Returning to Mark and Martin, it is clear that not even a journey to the pestilential heart of the United States represented by Eden is enough to overcome Martin and Mark's status as outsiders and allow them to reach any final conclusions about the American nation or character. This is not to suggest that neither of this pair offer definite opinions of United States and its citizens. What becomes clear on closer inspection is that even those opinions fail to erase the tensions of ambiguity and ambivalence that exist between the United States as an actual place and an imaginary, "amputated America" in much the same way that Martin's countering of Tapley's American eagle prevents it from being nothing more than an unmoving condemnation of the United States.

This same ambivalence, I would like to suggest, helped capture Dickens's imagination, causing him to visit the United states in the belief, as one critic has written that he was "a democrat, possibly even a republican, for whom America must indeed be the land of freedom, and thus politically and morally exhilarating."[25] It also caused the novelist to opine in *American Notes*: "I cannot help doubting whether America, in her desire to shake off the absurdities and abuses of the old system, may not have gone too far into the opposite extreme."[26] The point I wish to make is not that Dickens was inconsistent in his approach to the United States, though he may have been, but that the status of the United States as a severed appendage left it open, as an imaginative construct, to a variety of interpretations. These interpretations range from America as a safety valve and land of freedom as in the case of Joe Willet, a haven from the law, as with Samuel Pickwick, and a potential land of material enrichment as in the case of the younger Martin Chuzzlewit. At the same time, America as an imaginative construct can encompass the violence of a war that results in Joe Willet's loss of limb, the slavery and violence railed against in the volume two, "Chapter the Ninth" of *American Notes* and a whole range of violent, dishonest and simply filthy characters and characteristics found in *Martin Chuzzlewit* and elsewhere.[27]

Hannibal Chollop in *Martin Chuzzlewit* provides a final, striking example of the extent to which, from *Barnaby Rudge* onward, Dickens's imaginary republic contained a wide range of such impulses and possibilities. Encountered by Mark and Martin in Eden, readers are informed of Chollop that:

> He was usually described by his friends in the South and West, as 'a splendid sample of our na-tive raw material, Sir,' and was much esteemed for his devotion to rational Liberty; for the better propagation whereof he usually carried a brace of revolving-pistols in his coat pocket, with seven barrels apiece.[28]

Not only does this passage clearly posit Chollop's status as an exemplary even archetypal American of some sort through the twanging intonations of his friends' speech, it also makes clear the contradiction inherent in his position. Revolving pistols may have their uses, but the support of rational Liberty seems unlikely to be foremost among them. Chollop, it appears, not only has a propensity to violence as signalled by his arsenal, but also manages to justify that propensity through claiming devotion to some facet of an idealised image of America. The fact that this character justifies himself in such a way while earning the respect and even admiration of his countrymen for doing so indicates a self-serving and hypocritical deployment of American ideals. At the same time, this excerpt does not attack those ideals in and of themselves; the misappropriation of the concept rather than rational Liberty itself is what is frowned upon here.

More importantly, by highlighting the easy exploitation of the discrepancy between actions in reality and admirable, if ethereal, ideals Dickens uses Chollop to offer a final explanation for how his "amputated America" can include such apparently opposing impulses. What Chollop's self-serving deployment of American rhetoric illustrates is that for Dickens, and likely for other as well, the United States occupied a unique position at the point of intersection between the imaginary and the ideal, the actual and the imperfect. Nor is Chollop the only illustration of this in the novelist's work. Joe's and Martin's personal encounters with the country illustrate this same process of intersection, albeit to a less pronounced degree, as does the despair Dickens gave voice to when he failed to find his imaginary republic. Once Dickens's realisation of America's position astride a boundary between the imaginary and the real is recognised his continual grappling with a variety of imaginative renderings and values of the United States becomes more understandable than it might otherwise be. Though Dickens may not have recognised the final source of his amputated America's ambiguity when he gave Joe Willet and his missing arm the symbolic value that he did, he had clearly taken a first step toward such a recognition in his refusal to depict either the United States or its relationship with Britain in final, judgmental terms. Dickens may have seen much to cause him despair when he considered the United States causing him to write to one correspondent: " [. . .] I do fear that the heaviest blow ever dealt at liberty will be dealt by this country, in the failure of its example to the earth."[29] Even in the midst of such despondency, however, Dickens did not reject a more hopeful, regenerative role as a possible one for the United States, even if only as an example to the rest of the world.

Finally, this concept of an ambivalent, "amputated America" was not the sole imaginative preserve of this Victorian novelist. As his letters and other writings suggest Dickens was aware of and influenced by a variety of views of America current at the time, many of them originating in the United States itself. Robert Heilman, for example has argued that *Martin Chuzzlewit* "was not an initiator" of fictional renderings of America as something less than a utopia confirming that, at the very least, Dickens's range of responses to America as a construct and a place were neither precedents nor entirely unique.[30] Dickens did not create his vision of an "amputated America" in an imaginative vacuum and while through the figure of Joe Willet he may have given that image an unprecedented voice and power, it seems likely that both through the novelist's work, and through the lenses of historical and social debate available at the time, that the United States as a severed appendage and ambivalent locus of a variety of forces of good and ill would have been potentially available to other British subjects.

---

[1] Charles Dickens to W.C. Macready, 22 March 1842, in *The Letters of Charles Dickens*, ed. Madeline House, Graham Storey and Kathleen Tillotson, (Oxford, 1974), 156.
[2]Harry Stone, "Dickens' Use of his American Experiences in Martin Chuzzlewit," *Proceedings of the Modern Language Association of America* 72.3 (1957), 464.
[3]Charles Dickens, *Barnaby Rudge: A Tale of the Riots of 'Eighty,* Penguin Classics (London, 2003), 258.
[4]Ibid., 254.
[5]Charles Dickens, *The Life and Adventures of Martin Chuzzlewit*, Penguin Classics, (London, 1999), 331.
[6] Charles Dickens, *The Posthumous Papers of the Pickwick Club,* Penguin Classics, (London, 2003), 604.
[7] Charles Dickens To Andrew Bell, 12 October 1841, in *The Letters of Charles Dickens*, ed. Madeline House and Graham Storey, The Pilgrim ed., Vol. 2 *1840-1841* (Oxford, 1969), 402.
[8] *Barnaby Rudge*, 260.
[9]*Martin Chuzzlewit*, 222.
[10]*Martin Chuzzlewit*, 234.
[11]Robert H. Fossum and John K Roth, "The American Dream," *BAAS Pamphlets in American Studies,* no. 6 (1981), 7.
[12]Ibid., 6.
[13]Ibid., 8.

[14] Malcolm Bradbuy, *Dangerous Pilgrimages: Trans-Atlantic Mythologies and the Novel*, (London, 1996), 106.
[15]*Martin Chuzzlewit*, 516.
[16]Ibid., 516.
[17]*Barnaby Rudge*, 601-602.
[18] Rod Mengham, "Authority and Rebellion," in *Charles Dickens, Writers and Their Work* (Tavistock, Devon, 2001), 4.
[19]Ibid., 37-38.
[20]*Barnaby Rudge*, 602.
[21]Ibid., 601.
[22]Ibid., 602.
[23]Ibid., 600.
[24] Ibid., 603.
[25] Sylvère Monod, *Martin Chuzzlewit* (London, 1985), 36.
[26] Charles Dickens, *American Notes For General Circulation*, Penguin Classics (London, 2000), 65.
[27]Ibid., 250-265.
[28]*Martin Chuzzlewit*, 492.
[29] Charles Dickens to John Forster 24 February 1842, in *Letters*, Vol. 3 *1842-1843*, 90.
[30]Robert Bechtold Heilman, *America in English Fiction 1700-1800: The Influences of the American Revolution* (Baton Rouge, 1937), 280.

CHAPTER SEVEN

THE SOUTH IN THE BRITISH POLITICAL MIND: DIPLOMACY AND THE SECTIONAL CRISIS

DAVID BROWN & DAVID WALLER[1]

When Britons thought about the United States in the mid-nineteenth century their concerns were very different from those common in the twentieth. American geopolitical power and cultural hegemony were not as yet significant beyond their own continent and, although the country was of great economic importance to Great Britain by the 1850s and the principal destination for her emigrants, political relations were only just recovering after forty years of mutual suspicion following the War of 1812.[2] Diplomatic relations with the United States were not as important to the British government as those with her European rivals and personal contacts between the governing élites of the two countries were spasmodic.[3] Nevertheless, there were deep historical connections between the two nations, albeit not always of the happiest kind, important to many on both sides of the Atlantic. America was, in numerous ways, not a 'foreign' country at all to Britons, so obvious were the ties of kith and kin, and reference to 'the mother country' in the USA was not uncommon.[4] The diplomatic settlement of longstanding issues such as the Canadian border (in 1846) seemed, by mid-century, to presage a more pacific era of cooperation.[5]

That the history of Anglo-American relations in the subsequent decade and a half was in fact anything but tranquil has been one of the more intractable, and controversial, conundrums of diplomatic scholarship. The American Civil War of 1861-65 was to see a serious breakdown that threatened war between Great Britain and the United States on more than one occasion, despite this being regarded as disastrous by political leaders on both sides. The origins of Anglo-American antagonism during the 1860s are manifold, but many writers have sought to ascribe it to a combination of American suspicion of British imperial politics, seeking to frustrate the Republic's "manifest destiny" to control the North American continent, and ingrained British antipathy towards American social and political values. What may be called the "traditional interpretation," first set out by Ephraim Adams and Donaldson Jordan and Edwin Pratt in the early twentieth century, paints the crisis in the Anglo-American relationship as an accident waiting to happen.[6]

This conceptualisation of British views sees the country's ruling class, although anxious to preserve peaceful relations with the United States for economic reasons and to protect the largely indefensible Canada, as still not well disposed to its former colonies in the 1850s. American democracy was politically suspect, and potentially posed a revolutionary threat to the British Empire. Socially and culturally, Americans were regarded as rude, uneducated, provincial, capricious, grasping, excitable, and even child-like.[7] Thereafter during the war, the traditional interpretation runs, some in the Liberal government in London (most notoriously the Chancellor of the Exchequer, William Gladstone) even desired Southern independence, which would better suit Britain's strategic and political concerns. By contrast, the working class has been seen as much more pro-northern in its sympathies; and the victory of the Union in 1865 was a stimulus to parliamentary reform in Britain two years later. In sum, Britain's formal neutrality during the war and (an often inferred) latent sympathy for the cause of the Confederacy has been taken as evidence of a fundamental hostility towards an expansionist republican power that might in time threaten the British Empire–a power whose diminution, if not actual destruction, it was implied, might better serve British interests.[8]

This somewhat romanticised view has been steadily revised in the last thirty years since Mary Ellison published her pathbreaking work on the pro-Confederate sympathies of the Lancashire textile workers.[9] There is plenty of evidence to suggest that élite figures feared the revolutionary precedents which southern independence might establish and were relieved to see the North's eventual success, regarding it as the victory of legitimate government. Only the most progressive Liberals (such as John Bright) saw the United States as the pattern for Britain's own political development. Contemporary scholarship reveals a more complicated picture of a Britain that was deeply divided over the war at all levels of society. British opinion was highly nuanced and difficult to characterise as either pro- or anti-Confederate, as has been recently shown in great detail by, in particular, Richard Blackett.[10] The traditional picture of a conservative aristocracy that was pro-southern in sympathy and pro-Confederate in policy is now seen to be as inadequate as its counterpart depicting the British working class as united in support of a democratic Union waging a war of liberation against slavery and the 'slave power'.

Nor were British observers necessarily consistent in their attitudes: discomfiture at American actions that might harm British interests (and there were many) could be combined with regret that civil war threatened the future of a country whose progress was a prominent example of the superiority of British institutions and the British racial stock. The idea that racial ideology was fundamental to both British and American imperialism in the latter decades of the nineteenth century is now well established. It is possible, however, that an "Anglo-American political and cultural identity," as Martin Crawford puts it, was apparent by mid-century, reinforced by a shared ethnic heritage, with profound implications for the ways in which Britons viewed the conflict between North and South.[11] The sundering of the Union in 1861 might no longer be a cause for satisfaction at the humbling of an imperial rival but rather one of regret. Moreover there is evidence that the two contending parties were not viewed alike in the conflict, nor held equally responsible for the outbreak of war. Some British commentators blamed the lame efforts of James Buchanan's administration or the inherent weakness of the federal system, but a significant body of opinion attributed primary responsibility to the South. Most significantly, this was the position of at least one group of Britons (or representatives of the British Crown) who knew America intimately–Her Britannic Majesty's consuls.

An analysis of the correspondence of British consuls in the Southern states in the years leading to secession finds that Britons who had the closest connection to America in this era were sceptical of the claims of southern independence and strongly opposed to the continuation of slavery. Slavery emerges as the major impediment to cordial relations between Great Britain and the southern states, even prior to secession in 1861: racist legislation controlling the movement of free blacks in port cities was opposed by the British government; and British consuls (and ministers) rejected southern hopes of an alliance with British commercial interests because of their insistence upon preserving slavery. In time, of course, the abolition of slavery in the United States was to be seen as a great victory for a transatlantic antislavery movement that had drawn heavily on British funds and expertise. Moreover, supporters of parliamentary reform in Great Britain took inspiration from the triumph of American democracy over separatism and rebellion.[12] In the late 1850s and early 1860s, however, it was British consuls who arguably had the keenest insight into the momentous developments unfolding in the United States.

In the historiography of transatlantic diplomatic relations, a major element of the argument about British reservations towards America in the mid-nineteenth century has been the degree to which Her Majesty's Government was aware of political developments in the United States. Or unaware, one might better say, because it is usually argued that Britons misunderstood American federalism and were suspicious of democratic republicanism. In fact there is plenty of evidence to suggest otherwise. Despite their significantly different social and cultural backgrounds (different from Americans that is), the diplomatic correspondence of British foreign secretaries and ministers in Washington in the mid-nineteenth century reveals an acute awareness not only of the contours of

American foreign relations but also of internal partisan movements, the significance of electioneering, and the general political development of the American state.[13]

Her Majesty's Government was able to rely not only on its diplomats in Washington, but also on the highly experienced consular service throughout the United States. Consuls were stationed in major American ports and cities and, in addition to their routine duties related to commercial operations and the needs of British subjects resident in America, they provided social, political, and economic reports directly to the Foreign Office in London.[14] For example, when first appointed to the office at Mobile, covering "Alabama and the Floridas," Charles Tulin was invited to furnish reports on "commerce, navigation, agriculture or any other branch .... which may be interesting," for an annual salary of £450 and £100 expenses, although he was restricted from engaging in commercial activity.[15] A minority of consuls were naturalised foreigners (including some Americans), but those in the major commercial cities, including New York, Boston, New Orleans and Charleston, were experienced public officials of social standing in the local community.

The intelligence provided by consuls in the 1850s affords a valuable perspective on British opinion of the United States and challenges the preconception that the British élite remained rooted in a colonialist mentality towards America. It also demonstrates that a significant body of British opinion came to regard the differences between North and South as directly relevant to how the United States as a whole was to be understood. By the end of the Civil War, the South was firmly established in the British political mind as a region that was exceptional and, in many ways, deviant from norms and values common to both British and northern society.

Of the seven British consuls in the American South in 1860, Robert Bunch in Charleston and Edmund Molyneux in Savannah were particularly well placed to observe and report on the growing sectional crisis.[16] Britain's valuable commercial connection with the cotton economy had entailed extensive dealings with political and mercantile interests in these districts; and a history of assisting black British subjects who were adversely affected by racial legislation had given the consuls a detailed insight into southern politics. Moreover, Savannah (to an extent) but particularly Charleston, were hotbeds of secession in which radical fire eaters played a leading role in taking the southern states out of the Union. Consul reports to the Foreign Office in London (and occasionally to the British Legation in Washington, D.C.) are uniquely rich in their observations of, and comments about, the break up of the American nation.

In general, British consuls throughout the United States were impressed by the dynamism and economic potential of the young republic, if not by its social customs, but they perceived major differences between regions. The North and the West were seen as the lands of progress in comparison to a South dependent upon slave labour.[17] The differences between the sections extended to their politics as well. Lawlessness–or at least bias in the administration of justice (especially against British claimants)–and political corruption were seen as established features of U.S. public life throughout the country, in part because the educated social élite were "deprived of their natural influence" by the elective form of government.[18]

Southerners were reckoned especially prone to poor government. When commenting on the growing sectional crisis from the late 1850s onwards, Bunch and Molyneux argued that the South was thoroughly mistaken in its assumptions that separation would be achieved easily. Writing in October 1860, on the eve of a concerted attempt to force secession, Bunch suggested that "South Carolina has always been impulsive and hasty; always disposed to rebel against Federal dictation and restraint; but this very sensitiveness, whilst depriving it's [*sic*] threats of the weight which would attach to the utterances of a calmer people, may precipitate it into a course of action from which there will be no withdrawal."[19] A month later he acknowledged that an independent Confederacy was unlikely "without the antecedent horrors of a bloody civil war." Although southerners might have a legitimate grievance that the Republican Party was bent on destroying their "domestic institutions," as a people they (and especially those in South Carolina) were hasty, obsessed with their honour, and liable to act without restraint or much forethought.[20] As such they were, or could become, the antithesis of the gentlemanly conduct that élite Britons saw as fundamental to a rational political order.[21]

What is notable about the emerging British view of southern society–one for which the traditional interpretation of Anglo-American relations suggests they felt a close affinity–is its similarity to the opinion of the South held by northerners as the sectional crisis gripped the United States. As Susan-Mary Grant has shown, northern politicians and writers constructed in their public discourse an image of the South as a region not only substantially different from their own but also one that deviated from an idealised American norm.[22] It is not surprising that an ideology of northern nationalism should have been fundamental to the new Republican Party, but that the South was seen as different or even un-American by foreign observers too underlines how little sympathy the seceding states evoked abroad.

Throughout the early stages of the secession crisis, the British consuls comforted themselves that "conservative" (anti-secession) forces would prevail. After Abraham Lincoln's electoral victory in November 1860, however, it was acknowledged that unionism was a dead letter in the South, as much due to the influence of "demagogues" as any actual threats from the president-elect.[23] This was exasperating for some, who felt there were far more similarities than differences between the two sections. Bunch wondered whether it had actually been worth America breaking from its mother country, echoing a widespread view in Great Britain: "The alleged misdeeds of Great Britain, which led to the American Revolution, read but tamely when compared with those of which the North is now accused by the South." He found it difficult to accept the South's position on secession, believing that such "a Catalogue of grievances [was] not usually brought by a publick functionary of one portion against the Authorities and people of another section of the same Country."[24]

The consular interpretation of United States domestic politics therefore does not consist of a blanket condemnation of American republicanism. It was rather the excesses which could result from the democratic element or the machinations of individuals that were potentially dangerous.[25] Moreover, there was a recognition that southern opinion was hypocritical in its rejection of Lincoln's election: as Bunch suggested to Foreign Secretary Lord John Russell in October 1860, "To one who might be disposed to think that the practice of the people of the United States would be in accordance with the theory of their Constitution and Government, it must appear strange that a President who, if elected at all, will undoubtedly be chosen by a majority of the States, or, in other words, by the will of the People, expressed through it's [*sic*] legitimate channels, should fail to command support and obedience of the minority, who may have been disappointed in the success of their candidate. But in the United States, as elsewhere, profession and practice do not always agree; a truth of which, I much fear, the world is about to receive a startling demonstration in the present case."[26]

Consular evidence suggests a strong repugnance towards slavery and consequent preference for the North. Southern consuls like Molyneux and Bunch tended to view the South in a slightly patronising fashion, in places making judgements like a parent to a child, which of course was a charge made of the British attitude toward the United States in general. Contemporary Americans thought "British observers of the United States were ignorant, superficial, and suffused with a basic hatred of American democracy,"[27] although those observers were often quite capable of distinguishing between the political cultures and socio-economic conditions of the North and the South.[28] Bunch regarded the South much like an errant waif who refused to see the error of his ways by insisting upon perpetuating slavery. The views of southern consuls were influenced by their experience of dealing with southern state laws protecting slavery by regulating the behaviour of free blacks, which had significant consequences for black British subjects. Of particular importance in this respect were the Colored Seamen's Laws, which authorised the incarceration of black sailors arriving in southern ports on board visiting vessels. This adversely affected British sailors from the Caribbean. British opposition to such legislation was sometimes motivated more by pragmatic than moral concerns. Consul William Mure in New Orleans, for example, thought it potentially an "injury to the commerce of the City" which had damaging consequences for British business.[29] That said, all consuls seemingly rejected southern fears about the threat posed by free blacks and did their utmost to protect the rights of black British sailors. The government in London was at pains to stress that all Britons enjoyed the Crown's protection (which in the case of these sailors meant the assistance of the consular authorities in the American South) whatever their race or colour, a position rigorously upheld in the consular reports.

Given Richard Blackett's recent emphasis upon the importance of an antislavery "residue" within Britain in the 1850s, this is a particularly important point. Great Britain presented itself as the world's leading antislavery power in the immediate aftermath of emancipation in the British Empire in 1833, but historians have tended to depict this commitment on the wane in the middle decades of the nineteenth century. Frederick Douglass famously noted a rising tide of racism in Great Britain in between two visits in 1845 and 1859, finding "American prejudice [on the] streets of Liverpool and in nearly all...commercial towns."[30] Other historians have suggested that Britons had grown weary of the slavery issue, disillusioned with the effects of emancipation in the West Indies, and/or persuaded by the racist arguments of former Caribbean planters.[31] While to an extent there was a retreat from antislavery commitment, consul papers tend to support Blackett's contention that "there is ample evidence that a strong undercurrent of abolitionist sentiment continued to exist."[32]

Instructions issued to Consul Mure at New Orleans concerning the Colored Seamen's Laws were unequivocal: "Her Majesty's Government have been much disappointed by the determination of the Legislature of Louisiana to wrought a system which is no less discreditable to itself, than it is injurious to the persons whom it immediately affects, and they are at a loss to understand the grounds on which this harsh procedure has been adopted." The Foreign Office urged the consul to ensure that "better counsels will presently prevail" and ordered him to take whatever steps necessary to address the situation, including pressing any case through the federal courts. Louisiana, it was suggested, was not only "disregarding the rights of foreign nations" and violating the provisions of the United States Constitution but also ignoring "the common dictates of humanity."[33] In South Carolina Bunch wrote to a member of the state's House of Representatives, Colonel J. Harleston Read, urging better treatment of "my unfortunate Countrymen [who] were imprisoned for the offence of being what God made them" on the grounds that it would be in the South's own interests. "I cannot keep thinking that if any of the plans should be realized which have for years occupied the minds of Southern Statesmen, if the 'irrepressible conflict,' so loudly proclaimed, should ever lead to complications which it is easier to imagine than to describe, it would be wiser on the part of South Carolina to make a friend, rather than an enemy, of an old ally."[34] Perhaps the consuls' close proximity to the "peculiar institution" heightened their antislavery instincts, but communications they received from the Foreign Office were equally antislavery in intent.

For those British consuls below the Mason-Dixon Line, the issue of slavery became acute as the secession crisis deepened. In December 1860, Bunch in Charleston was approached by a representative of the state government, former U.S. Senator Robert Barnwell Rhett, editor of the Charleston *Mercury* and notorious fire-eater, who sounded out the consul on whether South Carolina could expect British support for her independence. Bunch was at pains to point out that, whatever arguments Rhett might make about the commercial or strategic advantages an independent southern nation could offer Great Britain, his country would have no sympathy for a regime which practised slavery and certainly could not entertain an alliance with any that considered reviving the slave trade "which Great Britain viewed with horror." Moreover, any southern hopes that other European states might step in where Britain feared to tread were undoubtedly ill-founded. "The universal detestation of the African Slave Trade felt by all civilized people" should make Rhett and his fellow secessionists cautious, Bunch stressed. They should "not forget that nearly all the Powers of Europe were bound by Treaty to repress it, and that it was hardly likely that they would tolerate in one Nation, for the sake of commercial gain, that which they had systematically and continually reprobated in others."[35] In taking this stance, Bunch was doing no more than repeating the view held in the British Legation in Washington and indeed in the highest circles in London.[36] The British Minister Lord Lyons, writing to Lord John Russell in the same month, repeated Bunch's sentiment almost verbatim: "It must ever be repugnant to our feelings to be in intimate relations with a Confederation formed on the avowed principle of perpetuating, if not extending, Slavery."[37]

As the war progressed, British views of the North were to become more favourable as opinion of the South declined. This was a quite remarkable development given the serious disputes Great Britain had with the Union during the course of the conflict, disputes such as the *Trent* affair of 1861 and the crisis over the Laird rams two years later that nearly led to hostilities between the two powers despite Great Britain's formal neutrality.[38] Part of the explanation lies in the adverse treatment that individual British subjects received in the Confederacy as its military and economic situation worsened. Even before the outbreak of hostilities, the consul in Charleston was approached by British subjects who were being forced to join the state militias. The contemporary law of nations permitted enforced service by foreign subjects if its object was the preservation of domestic order, but Bunch denied that this was the primary purpose of these patrols: ". . . the 'Patrol' alluded to, is in reality a Police force, which is charged more specifically with the supervision of Slaves, upon whom it may inflict summary chastisement for absence from their Owner's [*sic*] plantation and similar offences, without legal trial. It can hardly be contended that a British Subject, who is not a Citizen of the United States, should be compelled to participate in such acts of inhumanity."[39] Britishness, in the consul's eyes, was quite incompatible with inhuman conduct, so the South's violation of accepted standards of decency threatened to place it beyond the pale.

The situation for Britons worsened considerably, as it did for all foreign residents of the South, following the outbreak of war in April 1861. Military necessity required firstly the drafting of the state militias into the regular Confederate army and in due course the introduction of conscription in April 1862. In the chaos of war, British subjects, some long resident in the country, struggled to prove their nationality, and southern opinion turned heavily against Great Britain when it became clear that the European powers would not recognise the independency of the Confederacy. The British Minister in Washington, Lord Lyons, issued instructions to the southern consuls in November 1861 on the subject emphasising the injustice of any attempt to impress foreigners in a civil war in particular.[40] However the expulsion of the Crown's representatives from Confederate-controlled territories in mid-1863 removed the only means of official protection against ill-treatment that British subjects then enjoyed. Formally speaking, the President of the Confederate States Jefferson Davis revoked each consul's exequatur or legal authority from the host government recognising his status. The Confederacy had in fact long disputed the consuls' position because their exequaturs were originally issued by the U.S. Government.[41]

Official disputes in wartime are not unexpected and unremarkable in themselves, but Britain's alienation from the Confederacy was reinforced by the latter's status as a slave power. Formally speaking, the United States had been such before the secession of the slave states in 1861, but slavery was purely a sectional phenomenon and thus it was the South which came to be seen as the locus of inhumanity and even irrationality. In British eyes, slavery was an evil because (among other things) it debased the liberal political order that was the foundation of freedom. Lord Lyons observed as early as November 1860 that the popular election of Lincoln, which threatened abolition, had "carried men *of all classes* beyond the bounds of reason and common sense."[42] Attacks on abolitionists during the secession crisis provided "still further proof of the natural result of Slave Institutions in debasing and brutalizing the minds of

Slave owners."[43] Nothing that happened thereafter was to change this view in the official British mind.

Slavery, then, played a critical role in shaping the consuls' image of the South, as it did in a wider sense in influencing public opinion in Great Britain. As Duncan Andrew Campbell has shown, even those opinion formers who were sympathetic to the South, such as travel writers, recognised that slavery was an indelible stain on its reputation. The best that they could hope for was that independence might force the South's modernisation and economic diversification, perhaps in time causing slavery to wither away more rapidly than if the section continued to be joined with the North.[44] There is no evidence to suggest that Britons made any distinction between a political Confederacy that perhaps deserved independence as a matter of natural right–which they could support–and the South as a section of the United States that was culturally and economically backward in comparison with the North and with Great Britain–which they looked down on. To the nineteenth-century mind, there was a near symbiotic relation between a country's society and culture on the one hand and its institutions on the other and so the immorality of southern society fatally undermined its political cause. In any event, the existence of a significant pro-Confederate lobby in Parliament and among the British elite–one which supposedly influenced British government policy in favour of the South–is very largely a myth: most leading politicians, including many Conservatives, were either neutral in their sympathies or can be described as pro-northern.[45] Beyond the metropole, moreover, those who advocated southern independence often struggled to gain a hearing or were outnumbered by the supporters of the Union.[46]

Slavery was the anachronism that made the South different. James Spence, the Liverpool tin-plate merchant who was the driving force behind propaganda for the Southern cause in Great Britain during the war, believed that the Confederacy would have to initiate a policy of emancipation before it could possibly expect to gain general acceptance as a legitimate independent nation. Henry Hotze, the most important Confederate agent in England, disagreed with Spence. He promoted the idea of a hereditary and historical bond between Great Britain and the South: "Southern America, in manners, forms of speech, and habits of thought and business, resembled more old England, while Young England resembled more Northern America." The South was "proud of its closer affinity of blood to the British parent stock, than the North, with its mongrel compound of the surplus population of all the world." Unfortunately for Hotze's cause, the evidence suggests that few Britons shared this view. Confederate commissioner William Lowndes Yancey, sent to London in the summer of 1861 to seek recognition of his fledgling nation, was closer to the mark in observing that "antislavery sentiment was universal." If he had paid more attention to consuls stationed in the South, he would not have had to cross the Atlantic to reach this conclusion. Like the majority of Britons at home, they believed that slavery was an insurmountable stumbling block that forever condemned the South to be exceptional and, worse, unacceptable.[47]

Between 1859 and 1863, the course of southern secession fundamentally damaged, and finally terminated, Great Britain's official relations with the southern states. The failure of the Confederacy to achieve diplomatic recognition by the European powers was not, of course, determined exclusively by its own conduct, since in the case of Great Britain at least, the necessity to maintain peace with the considerably more powerful United States obliged the British to maintain their declared neutrality. Nevertheless, the sectional crisis itself was not perceived neutrally by Britons in America. The Crown's official representatives, both diplomatic and consular, were from the outset wary of southern claims to independence, often antagonised by its people's high-handed behaviour, and united in their condemnation of its slave system. Moreover, despite the very real crises in which Great Britain and the United States were embroiled during the war, these criticisms of society and politics were only rarely applied to the northern states.

In contradistinction to the traditional interpretation of Anglo-American relations, ruling circles in Great Britain were anything but pro-Confederate and indeed can be seen to have regarded the North as embodying more nearly the culturally and ethnically-based attributes of liberal civilisation that the British saw in their own system, however much tarnished in practice these were by the realities of war. The victory of the Union in the American Civil War, far from representing a disaster for the British Empire, was seen as the only legitimate outcome, even by many political conservatives, and even for some as a vindication of "Anglo-Saxon" civilisation. For as Consul James had informed the Foreign Office, writing from Boston in 1851, "This is a very wonderful country, and no Englishman that I know has done justice to it .... amidst this chaos of democracy, principles of great value and importance are slowly evolving themselves, and there are virtues beneath the surface of society."[48]

---

[1] The authors thank Martin Crawford and Robert Cook for their valuable comments on this essay. David Brown also gratefully acknowledges the support of a British Academy Overseas Conference Grant (OCG-36523) which enabled him to present an earlier version of this paper at the 51st International Congress of Americanists in Santiago, Chile, in July 2003.

[2] Martin Crawford, *The Anglo-American Crisis of the Mid-Nineteenth Century: The Times and America, 1850-1862* (Athens, Ga., 1987), 8. Bernard Porter, *Absent-Minded Imperialists: Empire, Society, and Culture in Britain* (Oxford, 2004), 26-27.

[3] The British legation in Washington D.C. was headed by a Minister and not yet an

Ambassador (as it would be from 1893).

[4] The years 1852-62 actually constitute the only period between 1850 and 1890 when the United States was *not* the principal destination for emigrants from England and Wales (in part because of the war): Charlotte Erickson, *Leaving England: Essays on British Emigration in the Nineteenth Century* (Ithaca, N.Y., 1994), 60.

[5] Crawford, *The Anglo-American Crisis*, 6-7.

[6] Ephraim Douglass Adams, *Great Britain and the American Civil War* (London, 1925) and Donaldson Jordan and Edwin Pratt, *Europe and the American Civil War* (Boston, 1931).

[7] Martin Crawford, "British Travellers and the Anglo-American Relationship in the 1850s," *Journal of American Studies* 12.2 (August 1978), 203-219.

[8] The persistence of this interpretation is traced in Duncan Andrew Campbell, *English Public Opinion and the American Civil War* (Woodbridge, Suffolk, 2003), 2-9.

[9] Mary Ellison, *Support for Secession: Lancashire and the American Civil War* (Chicago, 1972).

[10] R. J. M Blackett, *Divided Hearts: Britain and the American Civil War* (Baton Rouge, 2001).

[11] Paul Kramer, "Empires, Exceptions, and Anglo-Saxons: Race and Rule between the British and United States Empires, 1880-1910," *Journal of American History* 88 (March 2002), 1315-53. Crawford, *The Anglo-American Crisis*, 9.

[12] Campbell, *English Public Opinion*, 1-15.

[13] James J. Barnes and Patience J. Barnes, *Private and Confidential: Letters from British Ministers in Washington to The Foreign Secretaries in London, 1844-67* (Selinsgrove, Penn., 1993) and *The American Civil War through British Eyes*, three vols. (Kent, Ohio, 2003 and 2005).

[14] Eugene H. Berwanger, *The British Foreign Service and the American Civil War* (Lexington, 1994), 4-5.

[15] Foreign Office to Tulin: U.K. Public Record Office (PRO) Foreign Office Series 5 [hereafter FO 5] 699, p 55; 12 May 1858.

[16] British consuls in the South were stationed at Baltimore, Charleston, Savannah, Mobile, New Orleans, Galveston and (the only one not a port city) Richmond.

[17] Laura A. White, "The United States in the 1850s as Seen by British Consuls," *Mississippi Valley Historical Review* 19.4 (1933) 509-536, and "The South in the 1850s as Seen by British Consuls," *Journal of Southern History* 1.1 (1935), 29-48.

[18] White, "The United States in the 1850s," 524-526. According to Consul James in Norfolk, Virginia, no less a figure than U.S. Chief Justice Roger B. Taney was also critical of political democracy for the same reasons.

[19] Bunch to Foreign Office: PRO FO 5/745 p 165; 20 October 1860.

[20] Bunch to Foreign Office: PRO FO 5/745 p 220; 29 November 1860.

[21] On the connection between liberal political principles and ideas of racial Anglo-Saxonism in the nineteenth century see Porter, *Absent-Minded Imperialists*, 70-71.

[22] Susan-Mary Grant, *North Over South: Northern Nationalism and American Identity in the Antebellum Era* (Lawrence, 2000).

[23] Vice-Consul MacRae at Wilmington, North Carolina to Bunch: PRO FO 5/745 p 265; 24 December 1860.

[24] Bunch to Foreign Office: PRO FO 5/745 p 200; 24 November 1860.
[25] During the initial crisis (especially over the blockading of Southern ports and its effect upon Anglo-French commerce, in mid-1861) the British Minister in Washington, Lord Lyons, informed the Foreign Secretary Lord John Russell that threats of war against Great Britain were more the product of the personal opinions of Secretary of State William Seward than the considered policy of President Lincoln's administration: Barnes and Barnes, *Private and Confidential*, 251-252.
[26] Bunch to Foreign Office: FO 5/744 p 165; 20 October 1860.
[27] Crawford, "British Travellers," 203-204.
[28] Campbell, *English Public Opinion*, 117.
[29] Mure to Foreign Office: PRO FO 5/744 p 15; 16 February 1860.
[30] Frederick Douglass, cited in Blackett, *Divided Hearts*, 37. Blackett lists similar reactions by African Americans on p. 40.
[31] Douglas A. Lorimer, *Colour, Class and the Victorians: English Attitudes to the Negro in the Mid-Nineteenth Century* (Leicester, 1978), esp. 123-130. British Home Secretary Sir George Cornewall Lewis, writing in the *Edinburgh Review* in April 1861, "did not think the South could be expected to commit economic suicide, when British experience in the West Indies had shown that a free negro workforce was an unsatisfactory substitute for slavery"; but the argument of Lewis's article was that slavery was nevertheless a great evil (cited in E.D. Steele, *Palmerston and Liberalism, 1855-1865* (Cambridge, 1991), 301-302).
[32] Blackett, *Divided Hearts*, 54.
[33] Foreign Office to Mure: FO 5/721 p 1; 24 August 1859.
[34] Bunch to Foreign Office: FO 5/720 p 242-243; 28 November 1859.
[35] Bunch to Foreign Office; PRO FO 5/745 p 229; 5 December 1860.
[36] Steele, *Palmerston and Liberalism,* 296.
[37] Lyons to Russell, Dispatch 317, 18 December 1860; Barnes and Barnes, *The American Civil War through British Eyes*, Volume 1, 12.
[38] Howard Jones, *Union in Peril: The Crisis over British Intervention in the Civil War* (Chapel Hill, 1992), 80-99; Dean B. Mahin, *One War at a Time: The International Dimensions of the American Civil War* (Washington, D.C., 1999), 174-184.
[39] Bunch to Foreign Office: FO 5/745 pp 137-141; 22 June 1860.
[40] Lyons to Russell, Dispatch 660, 18 November 1861; Barnes and Barnes, *The American Civil War through British Eyes*, Volume 1, 212.
[41] Berwanger, *The British Foreign Service*, 92-96 and 108-112. Consul Moore at Richmond was the first to be expelled in June 1863; Consul Bunch had already left Charleston (in February) on the Foreign Office's instructions because of Confederate antipathy towards him. Not all were actually deported: a few retired to private life *in situ* and thus were able, when their states of residence were occupied by Union forces towards the end of the war, to resume their functions as British consuls.
[42] Lyons to Russell, Dispatch 283, 12 November 1860 (emphasis supplied); Barnes and Barnes, *The American Civil War through British Eyes*, Volume 1, 2.
[43] Bunch to Foreign Office: PRO FO 5/745 p 203; 27 November 1860.
[44] Campbell, *English Public Opinion*, 117-124.
[45] Campbell, *English Public Opinion*, 134-192.

---

46 To cite one provincial example, the town of Northampton, which during the war had a direct economic connection with the South through its supply of boots for the Confederate army, was nevertheless not particularly fertile territory for the agents of the Southern Independence Association: by 1863 their meetings were attracting a negligible audience compared with the numbers who attended pro-Northern speeches. See David Waller, "Northampton and the American Civil War," *Northamptonshire Past and Present* 8.2 (1990-91), 137-153.

47 Henry Hotze, cited in R. J. M. Blackett, "British Views of the Confederacy," in Joseph P. Ward (ed.), *Britain and the American South* (Jackson, Miss., 2003), 143; William Lowndes Yancey, cited in Crawford, *The Anglo-American Crisis*, 173, n.10.

48 White, "The United States in the 1850s," 535. Campbell makes the important point that perhaps the majority of Britons nevertheless did not think highly of either side in the conflict, not least because it had resulted in over 600,000 dead in the course of four years (*English Public Opinion*, 246).

CHAPTER EIGHT

FROM BARNUM TO 'ORGANIZATION MAN': IMAGES OF "AMERICA" IN THE BRITISH ADVERTISING DISCOURSE, 1850S – 1950S

STEFAN SCHWARZKOPF

The Transatlantic 'puff', compared with the European, is as Boreas to a breeze.[1]

I. Introduction

Visions of dexterous American advertisers and salesmen commercializing trans-Atlantic culture featured prominently in British perceptions of the young nation across the Atlantic. In the late nineteenth and early twentieth century, these images were created and communicated for different reasons by different groups within society. British advertising practitioners asserted their own professional identity and their competitive position against a perceived threat from American rivals by resorting to ideas of "essential" cultural differences and the rhetoric of national, cultural identity. This professional group projected its assumed "Englishness" against changing images of the "American advertising man." Until the First World War, American advertising and its alleged excessiveness often provoked moral outrage, rejection or condescending reactions among British observers. By the late 1930s, these perceptions had given way to images of the American "super salesman" as a highly specialised expert of middle class lifestyles. After World War II, the widely recognised efficiency and global success of American advertising became the epitome for the final victory of the "organization man" (W. Whyte) over the enlightened "gentlemanliness" of British advertising men.

II. "This is an age of advertising": the Barnum circus descends on Britain

From Charles Dickens to J. B. Priestley, British literary figures have interpreted their respective eras as the "age of advertising."[2] Nineteenth-century British journals and newspapers in particular voiced unease about what they interpreted as a change towards a culture in which both marketers and politicians only had to shout loud enough to sell anything. By the 1880s, this assessment was frequently made by political and business correspondents of the London *Times*, who often quoted the case of the United States as an abhorrent example for the influence of advertising on public and political life.[3] The impact of one of the greatest representatives of American salesmanship, Phineas Taylor Barnum, on the making of late nineteenth-century British attitudes towards advertising in general can not be overestimated and warrants further attention.

The Connecticut-born storekeeper, journalist, publisher and showman Barnum made a name for himself in the 1830s and 1840s for his travelling circuses, where he would exhibit the African-American slave woman Joice Heth, claiming that she was over one hundred and sixty years old and had been the nurse of George Washington; the "Fiji Mermaid," which was most likely the body of a fish and a baby monkey stitched together; the 33-inches midget Charles Stratton, celebrated as "General Tom Thumb" and other "human curiosities" whom Barnum presented ostensibly for their educational and scientific value. After his purchase of "Scudder's American Museum" in New York in 1841, Barnum became the most successful American showmen of the mid- and late-nineteenth century. At the peak of his enterprising adventures he engaged the Swedish singer Jenny Lind in 1850 to sing in America at $1,000 a night for one hundred and fifty nights. The difference Barnum made to American and European advertising practices resulted from his skilled management of the press and his ability to turn his exhibited products into "brands." Weeks before Barnum's circus visited a town, local newspapers would be filled with stories launched by Barnum himself, praising the coming excitement. Barnum thus developed the idea of using the press as public relations medium. Moreover, Barnum's strategy relied on adding sensational value to his "products": rather than advertising Jenny Lind merely as a world-class singer, he built her up as the "Swedish nightingale." Barnum's fame and success indeed came to define "American showmanship" and the cliché of American advertising as a form of "puffery": the trumped-up and noisy but hollow celebration of cheap sensationalism that undermined the moral fibre of society.[4]

In 1844/45, Barnum took young "General Tom Thumb" on a tour through the United Kingdom and continental Europe. On his tours in England, "Tom Thumb" appeared twice before Queen Victoria. Barnum's shows in Britain in 1844/45, 1857 and 1898 had a considerable impact on how British society perceived advertising and American advertising in particular. The journal *Punch* for example voiced strong disapproval of Barnum's shows and his advertising methods, such as the cluttering of exhibition theatres with degrading and tasteless bill posters. The discussion about Barnum's promotional practices in *Punch* reveals that debates about advertising were framed by contemporary concepts of trans-Atlantic political culture. In the eyes of American commentators, Barnum's shows exposed the hypocrisy of European kingships: while Europe's crowned heads spent fortunes on seeing the "loathsome dwarf," Europe's "men of genius" were left to starve in their garrets.[5] The journal *Punch* on the other hand argued that the only activity Barnum was engaged in was the "worship of the Almighty Dollar" and the fate of one of Barnum's most famous attractions, the "White Elephant," was likened to that of the descendants of Gold Coast Princes and African Kings, who had to blacken the shoes of the "free Republicans."[6] The London *Times*, compared Barnum to the famous trickster in Spanish literature Lazarillo de Tormes and concluded in 1855:

> Thus he may be presumed to have graduated in that modern art of 'puff' of which he afterwards became the most accomplished known practitioner. The Transatlantic 'puff', compared to the European, is as Boreas to a breeze, or P. T. Barnum to any other showman that ever advertised his wares.[7]

While Barnum's stunts came generally to be equated with "American" styles of advertising, the moral outrage shown by *The Times* and *Punch*, however, was not necessarily shared by the majority of British society, not even by the upper echelons of society at the Royal Court in Victorian London. Regarded by some as a social and cultural threat to the stability of traditional European societies, Barnum's touring shows lured the uneducated poor as well as the educated and well-mannered middle and upper classes, which showed great desire to see the giants and dwarfs, the albinos, "red Indians" and "negro warriors," the mermaids, jugglers, woolly horses and white elephants. Barnum's seductive game consisted of his well developed understanding of the desires of all classes which he fed with sensationalist advertisements, flaming posters and transparencies, handbills, flags, banners, bands, and electric signs. On his several tours in Britain, Barnum's midget "Tom Thumb" amused the Liverpool underclass as well as the Royal household. When little Tom first met Queen Victoria in March 1844, he greeted the Prince of Wales with "How are you Prince" and intoned the "Yankee Doodle" in front of the Queen, much to the embarrassment of the Lord-in-Waiting but also to the great enjoyment of the Queen and her entourage.[8]

The undercurrents of cultural democratisation which characterised Barnum's shows and the new tools he developed to communicate with his consuming public through advertising were seen by Tory magazines as part of the generally degrading impact republican political systems had on their citizenry.[9] Some representatives of Britain's cultural and political establishment saw P. T. Barnum and his marketing campaigns as the vanguard of a coming age of mass democracy and its "dictatorship" of the susceptible many over the capable few. When news reached London of the remarkable success in the United States of Barnum's latest hype, the Swedish singer Jenny Lind, *The Times* reminded its readers of the nexus between Barnum's advertising and its potential to undermine the mental capacities of citizens in both republic and monarchy:

> It is the peculiar boast of the modern republic that the public opinion of her free and enlightened citizens reigns with undisputed and absolute sway. Eschewing the enormous faith of many made for one, she has adopted for herself the creed that the few are made for the many. On every subject, in every township throughout the States, the opinion of the majority is final, conclusive, and indisputable. … Any one impressed with these reflections must have perused with a painful interest the accounts which have from time to time appeared in this journal of the Lindomania in New York. It is humiliating to see a nation which boasts that it leads the van of human improvement so little capable of appreciating the relative dignity and merit of different talents and employments as to bow down in prostrate adoration at the feet of a woman who, after all, is merely a first-rate vocalist.

The article concluded:

> It is much to be feared that the same reckless system of exaggeration, the same vulgarity of means and littleness of ends is to be found in the Senate as in the orchestra. Who cannot see in the angry and inflated tone of American political controversy, and its constant straining after dramatical effect, the career of men to whom the most important measures, the most sacred interest and the most stirring appeals are matters of the same indifference as the comfort and quiet of Jenny Lind to Mr. Barnum when compared to the acquisition of a single cent?[10]

III. British attitudes towards American advertising practices

British attitudes towards modern American advertising practices at the end of the nineteenth century were deeply imbedded in the long-lasting impression

Barnum's sensationalism had left on the collective memory of his British audience. J. T. W. Mitchell in his Presidential Address to the Co-operative Congress in Rochdale in 1892 summarised these feelings when he identified advertising as the very heart of what he called the "Barnum wickedness of the competitive world."[11] Still one century after Barnum's first visit to Britain in 1844, his name was a centre-point for all those who saw the core of "American values" and the "American character" in commercial advertising and the ideology of consumerist waste of resources. In 1944, the British author Hamilton Basso drew a straight line between the age of Barnum and the sway modern advertising held over the twentieth-century "American dream" of consumer affluence:

> Out of a wedding of the circus and the New England commercial instinct was born a new and generic art which forever after was to be the most beguiling maidservant of business, teaching a nation to want things it did not need and throw away other things it could still use."[12]

This negative image of advertising was only intensified by a number of incidences at the beginning of the twentieth century. In summer 1900, at a time when the swindling excesses of Barnum and his business partners were still in everybody's memories, an American advertising agent representing "Quaker Oats" erected a billboard for the cereal on top of the White Cliffs of Dover to greet the incoming ships from Europe. The tremendous outrage this poster caused was voiced most effectively by the Society for the Checking of Abuses of Public Advertising (SCAPA), the secretary of which complained to *The Times*:

> One of the American food companies has got some one to erect high up on the cliffs two monster boards on which the name of their product is painted in letters that dominate and degrade the whole prospect. We ought perhaps to regard with a resentful sense of gratitude the feat of this remote Chicago firm in demonstrating to what lengths in the absence of rational restraint the instinct of vulgar competition can go in affronting the public eye.[13]

In order to avoid further damage being done to the image of advertising, a deputation of the London advertising trade approached the American firm responsible in order to have the billboard pulled down, yet without any success. It was only after Dover Corporation enacted a local Act of Parliament that the "Quaker Oats" poster was finally removed from its place on top of the cliffs. While much public outrage was expressed over the fact that advertising policy decided in America was able to sully a symbol of "English identity," there was no reference to another billboard, clearly visible in contemporary photographs of the site, advertising Pexton's, a drapery based in Dover.[14]

Only one year later, the same public scenario of Britain feeling besieged by American commercialism repeated itself. In 1901, James Buchanan Duke's American Tobacco Co. began to embark on an "invasion" of Britain by purchasing the Liverpool-based cigarette manufacturer Ogden's. This move sparked an unprecedented advertising war in the UK with British tobacco firms and American Tobacco stuck in a battle over promotional offers, price cuts and the "Britishness" of British cigarettes. American Tobacco continued to market Ogden's Guinea Gold cigarettes under the image of being a British cigarette. Advertisements for Ogden's either showed images of the British workforce at the Liverpool factory or a bulldog terrier defending Ogden's cigarettes as "British made by British Labour: Let 'em all come!" Other Ogden's advertisements warned smokers not to be gulled with imitations or to fall for the Anti-American propaganda of competitors.[15]

Ogden's and American Tobacco's largest rival in the United Kingdom, Godfrey Phillip's, retaliated with advertisements exposing Uncle Sam and his "American trust" behind the fake bulldog of Ogden's.[16] Another advertisement took on the image of the gull and showed John Bull discharging his gun at the American gull representing an "American trust."[17] Other advertisements by Godfrey Phillip's advertisements showed Uncle Sam plunging a dagger into the back of John Bull or a lion crouching defensively over a display of the firm's cigarettes and snarling at a top-hatted and star-spangled octopus in the sea. In the background of the advert stood the Statue of Liberty with the word "TRUSTS" radiating across the sky. In November and December 1901, London's newspapers were abound with American advertisements trying to assure British smokers of the higher quality of Ogden's Guinea Gold and the imported cigarette brand "Sweet Caporal." British advertisements retaliated by reminding smokers of the thousands of British workers whose employment was decided with every purchase made.[18]

In December 1901 eventually, the leading British firms reacted to the threat of the American Tobacco Company by forming a trade alliance, the Imperial Tobacco Co. In advertisements announcing the formation of Imperial Tobacco to the British public, the company again activated patriotic emotions and an anti-American rhetoric of foreign trusts and monopolies subjugating the free-born Englishman:

> Americans, whose markets are closed by prohibitive tariffs against British goods, have declared their intention of monopolizing the tobacco trade of this country. It is for the British public to decide whether British labour, capital and trade are to be subordinate to the American system of Trust Monopoly and all that is implied therein.[19]

The debates waged in contemporary publications about the undue influence of American advertising and marketing policies over the "British way of life" show the difficulties the social elites had in coming to terms with the rapid change of British society towards a modern, commercial mass democracy. In 1902, F A McKenzie sounded the alarm when he noted that

> The American invasion of Europe is no longer a matter of abstract discussions. It has touched Europe at a hundred points, and has affected no country so largely as our own. … America has invaded Europe not with armed men, but with manufactures products. Its leaders have been captains of industry and skilled financiers. No nation has felt the results of this invasion more than England. … From shaving soap to electric motors, and from tools to telephones, the American is clearing the field. … Our chemists' shops are full of Transatlantic drugs. Our bootmakers devote their windows to the finest manufactures from Boston, while our leading shopkeepers go across the Atlantic to learn the art of window dressing. … Our babies are fed on American foods, and our dead are buried in American coffins.[20]

As a result of such widespread fears, Britain's elites adopted a point-of-view in which all things modern in the world of media, marketing, advertising and publicity essentially had to be of American origin. The aggressive, populist political advertising campaigns by tariff reformers around 1910, for example, were decried by the old land-owning classes and traditional Free Traders as a form of "Americanising" of politics by the new "mercantile magnates."[21] The eleventh edition of the *Encyclopaedia Britannica* (1910) followed a similar course by attributing the development of modern newspapers and of modern printed advertising designs in Britain almost exclusively to the influence of American media. The article on "Advertisement" for example stated: "The ingenuity displayed in modern newspaper advertising is unquestionably due to American initiative." The article went on to explain that American newspaper editors printed headlines "as large as a man's hand," therefore US advertisers did not submit to a decorous code of conduct as adhered to by the British advertiser. Similarly, the article on "Newspapers" stated: "The modern impulse, culminating in England in the last decade of the 19th century in what was then called the 'New Journalism', was a direct product of American conditions and ways of life."[22]

The perceptions of American media and American advertising practices as modernising as well as potentially destructive powers were in line with the general tone of Anglo-American cultural exchange in the early twentieth century.[23] These trans-Atlantic perceptions are best reflected in the way contemporary literary accounts in the UK and the US interpreted the impact of American advertising practices on British society. Mark Twain's famously anti-

British satire *A Connecticut Yankee in King Arthur's Court* (1889) had an industrious American businessman using the methods of modern advertising in order "to clean up" the backward and inhumane class system of medieval Britain.[24] British authors on the other hand alerted their public to the menacing activities of American industries in Britain and the danger of British consumers being exploited by the aggressive marketing of American products. H G Wells' *Tono Bungay* (1908) for example painted the picture of American advertising methods as dishonest, dangerous, and utterly un-British. In Wells' satire of political corruption and the patent medicine system in the UK, the scallywag Edward Ponderevo, a Jew, is using the new American type of sloganeering in order to sell "slightly injurious rubbish at one-and-three halfpence and two-and-nine a bottle, including government stamp." Ponderevo's eager focus on the *sales* of pills rather than their product *quality* is accompanied by slogans such as "Not a drug, but a Live American Remedy."[25]

The interwars years were in a sense prejudiced in the same way. George Orwell, known for his literary defences of the individual against the mind-controlling super state, also saw the arrival of American advertising and consumer culture as the first step towards the final decline of human dignity. Orwell's 1936 novel *Keep the Aspidistra flying* introduces the reader to the inner side of an all around "modern" advertising agency-the New Albion-and the mental habits of the people inhabiting this world:

> The interesting thing about the New Albion was that it was so completely modern in spirit. There was not a single soul in the firm who was not perfectly well aware that publicity—advertising—is the dirtiest ramp that capitalism has yet produced. In the red lead firm there had still lingered certain notions of commercial honour and usefulness. But such things would have been laughed at in the New Albion. Most of the employees were the hard-boiled, Americanised, go-getting type—the type to whom nothing in the world is sacred, except money. They had their cynical code worked out. The public are swine, advertising is the rattling of a stick inside a swill-bucket.[26]

IV. The perception of American advertising among British advertising men

Criticism of ostentatious American advertising practices in the United Kingdom was anything but confined to the literary elite, the Left or members of the old political establishment. By the late nineteenth century, British advertising agents, *themselves* at the centre of much controversy, had started observing the professionalised, more powerful American advertising industry with a mix of contempt and admiration, envy and anxiety, too. This culture of observation, self-comparison and self-assertion emerged within a fixed set of social encounters, such as trans-Atlantic travels and conferences, and found expression mostly in British advertising trade journals. British advertising men used these encounters in order to negotiate the extent to which they saw an "American influence" on their industry as acceptable.

From the late nineteenth century onwards, the presence of American advertising in London grew steadily, if slowly. The first advertisement offices representing American newspapers appeared in London in the 1870s and 1880s. One of the first American advertising agents was Paul E. Derrick, who moved to London in 1896 representing American clients such as Quaker Oats, Regal Boots and Glycerine Tar-Soap.[27] In 1905, the American agency Dorland opened its offices in London. In 1919, J. Walter Thompson and Erwin, Wasey & Co. followed. In 1922, Lord & Thomas and McCann-Erickson became the next American multinational advertising agencies which moved to London. Until well into the 1930s, American agencies in London relied on a very small number of American clients which provided the mainstay of turnover needed to run a profitable business. While large British agencies in London usually counted more than hundred employees, J. Walter Thompson's London office staff at Bush House remained limited to 14 between 1919 and the middle of the 1920s. Thus, the actual presence of American agencies and the palpable influence of Americans on British advertising before mid-1920s was rather limited.[28]

In the first decade of the twentieth century, however, British advertising agents began to realise that the higher social recognition advertising enjoyed in America as well as its comparatively higher profitability in the US meant that the American advertising industry was about to undergo revolutionary changes in the professional quality of its work. The most important marker for these changes was the earlier development of advertising professional bodies and trade associations in the US. In 1904, American agents had founded the National Federation of Advertising Clubs of America, which in 1914 was renamed into the Associated Advertising Clubs of the World (AACW).[29] This body held annual conventions, which since 1911 attracted large British parties of delegates who returned home deeply impressed by the professional standards of their American colleagues and competitors.[30] The British trade journal *Advertising World* for example described the 1911 Boston Convention as "a revelation to those to whom advertisement still connotes bluff, mendacity, the methods of cheapjack and the wiles of the swindling showman."[31] Most importantly, the American advertising profession itself established an organisational structure which helped demolish the Barnum-image that had prevented the US agencies from communicating themselves as leading the "van of human progress." The American conventions of the AACW also brought about the "Truth in Advertising" movement, which served as an emotional rallying point for the American advertising industry and its attempts to modernise the image of advertising and publicity.[32]

Throughout the first three decades of the twentieth century, British advertising men tried to emulate the successful professionalisation strategies they had seen at work in the United States. Through the organisation of large-scale exhibitions, conventions and conferences in London, the British advertising industry not only followed the examples of the American advertising industry. The exhibitions and conventions staged in London since 1920 also showed the world that "America" had become an inspiring ideal to the British advertising industry. Beginning with the Advertising Exhibition staged at White City in February 1920, British advertising men organised a whole number of exhibitions and conventions designed to attract public attention to the new professional identity the industry tried to define for itself. This new identity of advertising as a developer of business and promoter of universal progress worthy of the best-educated young men relied on the presence of American representatives in order to authenticate the cultural and social aspirations of the young British advertising industry. Especially the "World Advertising Convention" held in July 1924 at London's Olympia became a stage for a level of Anglo-American fraternisation hitherto unseen and unheard of in the world of advertising. Newspapers and trade journals announced widely the arrival of a delegation of several thousands of American advertising men and women in London in summer 1924.[33]

For one week in July 1924, thousands of advertising men from all over the world dominated the Wembley exhibition ground, where the first Empire Exhibition took place simultaneously.[34] While virtually all modern nations had representatives at the Convention, the ideological horizon of it was clearly designed by an attempt to define "modern advertising" essentially as an Anglo-American project set out to enlighten the life of all members of society, spread the riches resulting from mass production, by that solving the problems of world trade and promoting the ideals of the world a peaceful commonwealth of nations.[35] At the opening session of the Convention at the Royal Albert Hall, greeting telegrams by King George V and President Calvin Coolidge were read out on a stage that was decorated with a silk model of the "Mayflower" ship. Amidst the cheers of the gathered advertising professionals, the Prince of Wales, Stanley Baldwin, Winston Churchill and the American Chairman of the AACW, Lou Holland, gave speeches and told the crowd that their business the was the "clarion of world trade," the greatest power to keep world peace and that they held the tools in their hands to restore the world's consuming power.[36]

The British advertising men listened to numerous speeches and reports of case studies prepared by their American counterparts, for example about the use of store newsletters and mailing lists in the States or about the service American agencies provided to their clients.[37] Other sections of the Convention dealt with the different structures of industrial administration in both countries and the impact of this for the existence of better market research facilities in North America as compared to Britain.[38] In order to publicly underline the stance both the British and the American professionals took regarding the Anglo-American project of a new century of advertising, highly symbolic gifts were exchanged at the Wembley Convention. These gifts ranged from state and city flags to a bronze statuette sculpted by the American artist Grace Pruden Neal. The bronze group showed two figures representing Columbia and Britannia in Greek dresses holding aloft the torch of "Truth," which since 1911 had been the symbol of the "Truth in Advertising" movement.[39]

One section within the British advertising industry which eagerly absorbed all new methods and techniques from the United States was the poster advertising industry. As early as 1900, letters and articles appeared in the outdoor advertising industry's trade journal *The Bill Poster* calling for a stronger orientation on trends in the United States. One letter in November 1900 for example complained that the American trade journal *Billposter*, published in New York, contained far more useful information for both agencies and clients and focused much stronger on enhancing the social role of the billposting industry.[40] In the struggle for a higher social recognition of advertising in society, the sheer beauty of colourful advertising posters was seen as a useful weapon against criticism. Here again, one British observer in 1925 found American posters more useful as they simply contained much more colours and warned his colleagues against condescension: "One of our greatest failings is to reject an idea on the basis that it is 'American,' sometimes when it isn't at all, and other times without giving it careful consideration before the emphatic 'No!'"[41]

Caricatures of American advertising men which appeared in British trade journals since the early 1920s began to show Americans as young, well-dressed, smart and slick professional office workers.[42] These men embodied a new managerial age descending on Britain and not any longer the preposterous, potbellied Barnum. Some British advertising men also began to engage in a frenzy of overrating all things American as the very shape of things to come. In some parts of the advertising industry, especially within the larger British agencies that had to compete directly with American agencies in London, "using USA methods" in market research became a byword for modern and successful advertising altogether.[43] In a similar vein, the advertising manager of Lyon's, W. Buchanan-Taylor, told his fellows and colleagues at the 1939

Advertising Convention in Blackpool that British advertising was "years behind" advertising in the US.[44]

This discourse of "America" as a promising ideal, however, did not necessarily affect all members of the British advertising industry. Voices could be heard within the profession which condemned this new influx of American efficiency, American designs and American sales methods. At a joint Anglo-American dinner held in January 1923 in preparation of the 1924 International Advertising Convention, the representative of the Association of British Advertising Agents, James Strong, declared that he looked forward to meeting his American colleagues as he was sure the British "...had something to teach the Americans." The trade journals *Advertiser's Weekly* quoted Strong further:

> He was not one of those who believed that good advertising originated in America. He believed that the English advertising agents knew something of their business, and he was sure when the Americans came over to London the advertising agents would give them something to carry back.[45]

One of the best-known and most outspoken British advertising men of his generation, Charles F. Higham, rejected the growing stress advertisers put on market research simply as an adoption of useless "American methods" and "decorative deception."[46]

Numerous voices in British trade journals also warned that "American" advertising and sales methods were not easily transferable due to the differences in the "buying psychology" of the British consumer.[47] In July 1927, the British advertising man Frith Milward for example complained about "these awful, high-powered, Americanised supersalesmen, so prevalent today." According to Milward, this new generation of "Americanised" salesmen tried to apply knockout advertising and sales methods they had seen at work in New York or Chicago but eventually failed, as the British consumer wanted to *buy* goods rather than *be sold*.[48] A similar chord was struck by a whole number of eminent British advertising agency managers in 1925. When asked by the trade journal *Advertising World* about their opinion regarding the increasing trend of US agencies marketing American goods in Britain, most advertising managers warned of the entirely different "buying psychologies" on the two sides of the Atlantic. William Crawford, owner of Crawford's Ltd., stated that nations had different "national" or "group minds," which foreign agencies simply would not understand. Albert Greenly, of Greenly's Ltd., described the British buying public as too cautious, too suspicious, for the typical American sales methods. In the same vein, Charles Higham and Ralph Winter Thomas attested that a great many American manufacturers had turned to British agencies after having found the "slangy" American advertising written in Chicago and New York for the British market a disastrous waste of advertising money.[49]

Some British advertising professionals visiting the US also returned with a reassuring sense of national and cultural differences between the English-speaking worlds on both sides of the Atlantic. Throughout the 1920s and 1930s, it almost became a ritual for British advertising managers to disembark Ocean liners from New York and announce that there was little the British had to learn from American advertising techniques. One such traveller was Charles Bronkhurst, advertisement director of the magazine *John Bull*. In his interview to the *Advertiser's Weekly* in July 1939, Bronkhurst stated that "his visit had shown him that America had little to teach us about advertising." He considered that "the make-up of the American newspaper was inferior to ours" and found that "British advertising copy was, in the main, equal to and in many instances superior to some of the American stuff."[50] Before the Second World War, thus, the terms "American" and "America" were also used by British advertising professionals as markers of difference and anchors of self-assertion in a world of constantly changing fashions and business fortunes.

V. Britain meets the Organization *Man*

The years after the Conservative recovery of power in 1951 saw an increasing influence of American advertising agencies in the United Kingdom. In fact, contemporary observers of the industry praised and loathed the era for the almost unstoppable rise and rise of American agencies in London, such as J. Walter Thompson, McCann-Erickson, Young & Rubicam, BBDO and Leo Burnett. These American-owned multinational advertising agencies embarked on a very aggressive expansion course of acquisitions, mergers and takeovers, to which even the largest British agencies fell victim.[51] While in 1937, only four American advertising agencies had branches outside the United States, by 1960, this number had risen to 36. Also by 1960, there were twelve American agencies in London which amongst them controlled some 30% of the aggregate advertising billings. In 1982, only four of the top twenty advertising agencies in the United Kingdom were still owned by British interests.[52] This rapid global expansion of American advertising became one of the main foci for the post-war European discourse of cultural and economic "Americanisation."[53]

Contemporary observers of Britain's "Americanisation" pointed at advertising campaigns for American cigarettes, cars, fashion items and movies as threats to the competitiveness of British creative and manufacturing industries.[54] This was made worse by the fact that due to the low output of British manufacturing and the strength of the Dollar in the 1950s the United States were seen as a market that was difficult to enter for the British exporter and his products. These circumstances posed challenges to the idea of advertising and marketing as a common Anglo-American project. British advertisers and exporters now viewed the United States as a different world and even demanded governmental support to help them understanding and overcoming these differences. Thus, the post-war British advertising discourse began to use "America" and "American" as a synonym for "the other" which needed explanation and cultural translation.[55]

An impressive source for this insistence on the "otherness" of American commercial cultures is the remarkable incoherence between the external and the internal communication of then largest British advertising agency London Press Exchange (LPE). In its communication to the public and to potential clients, LPE always stressed the fact that its organisational facilities and research services were on a par with what American agencies were offering to their clients. One advertisement issued by the LPE in 1960 claimed that the sheer breadth of different services the agency offered from its numerous offices around St. Martin's Lane and St. Martin's Square had turned this area of London into the "English Madison Avenue." The advertisement continued: "We have always admired American ideas about the proper scope of advertising agency service. In fact, we have not only admired it–since the 1920s we have been giving it to our own clients."[56] The image of an agency keeping up-to-date with trends in American advertising was underpinned by LPE's acquisition of the New York-based advertising and market research agency Robert Otto Inc. in 1962.[57]

The internal communication which took place within the agency at the same time, however, conveys the picture not of an advertising agency embracing the American advertising world but that of an agency asserting its "essential" difference from everything "America" stood for. In particular, the letters and travel reports which appeared in the LPE house magazine *In and out the Lane* speak a language far removed from the agency's official admiration of American advertising practice. One of these travel reports, appearing in 1959, tells the story of Angus Shearer's business trip "Among the Status Seekers."[58] The Englishman Shearer used his report to describe America as a foreign land were he was offered coffee instead of tea and where, outside New York, English visitors were still a novelty. Although he found that the English were generally welcome ("the uppercrust American is often a great Anglophile"), he also assured his colleagues back at home that the Americans were not in the least better at the job of advertising:

> They probably think they are. But they don't really know very much about the standards of advertising we have in this country and I think most of them would be quite surprised to see how advanced we are in many techniques.

The travel reports submitted by LPE staff to the house magazine also show that some people felt rejected by their American colleagues and business partners on the grounds of different notions of professional masculinity. American advertising executives placed great value on a methodical and sales-oriented approach as an attribute of the "manliness" of someone working within a corporate environment. English business men, on the other hand, ran into the danger of being perceived by their American counterparts as disorganised, "arty" and decadent for attitudes which would have been understood as part of a gentlemanly culture within the British context.[59] The indifferent politeness encountered by Shearer in 1959, for example, is entirely missing in a travel report that appeared in 1964. Here, an English LPE staff member travels to New York to visit a client but is told by the American: "Don't ever wear suede shoes to this office again. I regard them as a sign of English decadence." Life in the US was described by this Englishman as governed by the three principles of "conformity, simplicity, uniformity." New York subway trains were allegedly dirty even by London standards and altogether, money seemed the key to everything: "it breeds togetherness," not contempt as in England.[60]

A similarly "gendered" approach to the differences between American and British marketing and salesmanship cultures was used by an anonymous contributor to the government- and industry-sponsored market research journal *Markets and People* in 1951.[61] As a member of the British Export Trade Research Organisation (BETRO) who had just returned from London to New York, this contributor felt that the British business man was too naïve and "far too gentlemanly in his approach to selling" compared to the structured and disciplined sales organisation of most American companies. However, this tentativeness, he argued, "the almost feminine uncertainty" of the British manufacturer, could also be an advantage on the world markets in the light of the often misdirected zeal of the American school of salesmanship.[62]

The general picture of mutual "otherness" of English and American advertising cultures in the post-war period is confirmed by the experiences of the young Joan Bakewell, who later became a famous journalist and BBC presenter. Bakewell began her career as a copywriter for the American McCann-Erickson agency in London in the early 1950s. At McCann, her tasks included the "translation" of package inserts for American cosmetic and hygiene products such as "Tampax" from their American version into "proper English." According to Bakewell, American top managers would regularly fly in from New York in order to convene with local English managers. Due to their different dress codes, accents and habits, the "merrican" agency personnel would be looked at by young English staff as "aliens."[63]

At some moments, the "othering" of Americans and America by British advertising professionals took the shape almost of an ethnological discourse. In the early 1950s, a series of pamphlets by the Dollar Exports Board explained the characteristics of the American market to the British exporter in terms of a business and consumer ethnography.[64] The second brochure issued by the Board in 1950 on American advertising and sales promotion practices included a detailed drawing of a typical American High Street on which British consumer goods would have to compete. This drawing of an idealised heart of American post-war consumer modernity laid out how chain stores, variety stores, specialised shops, banks and recreation centres in an American town offered a great number of sales points and distribution channels.[65] The pamphlet's text explained:

> This is the shopping centre of a North American town. The branches of chain organisations are shown in red. The distributor's job is to keep goods flowing into the many varied retail outlets. The British exporter can get his goods into the flow by arousing consumer demand for them.

This pamphlet and a number of other publications by the Dollar Exports Board and its Advertising Advisory Council, the Federation of British Industries, the British Export Trade Research Organisation (BETRO), or the trade journal *Advertiser's Weekly* attempted to explain to British manufacturers how firmly the "habit of buying by brand" was enrooted in American consumer culture, how American advertising agencies helped the manufacturer in finding and interpreting vital market information, how American consumers deciphered advertisements and how the American manufacturer viewed advertising not as expenditure, but as investment.[66]

The plethora of organisations, publications and pamphlets which emerged in the late 1940s and early 1950s, designed to open up the horizon of the American consumer market for British products, reveals that the new post-war world of the super-efficient, affluent, sub-urban "organization man" was a world that needed to be explained to the British manufacturer by the American specialist. Yet at the same time, this network of organisations and their publications invested in a positive, pro-Atlantic mood similar to that communicated in the advertising conventions and exhibitions of the interwar years. One of the most important differences between the discourse in the interwar years and that of the post-war years is the recognition that the world of American large-scale retailing and national advertising campaigns was now open to the British manufacturer, too. Thus, while the post-war discourse asserted cultural differences, it also strengthened the idea of a common, trans-Atlantic commercial culture. Moreover, sociologists, writers and journalists on *both* sides of the Atlantic divide began to brand this culture as purpose-built by Madison Avenue, as dominated by the "man in the gray flannel suit," and in danger of being overpowered by the bureaucratizing, homogenizing activities of the "hidden persuaders."[67]

---

[1] 'The Life of T. Barnum', *The Times*, 5 January 1855, 10.

[2] J. B. Priestley is famously quoted: "Living in an age of advertisement, we are perpetually disillusioned. The perfect life is spread before us every day, but it changes and withers at a touch." Charles Dickens' observations of advertising in London's papers and on London's streets led him to conclude he was living in an age characterised entirely by the messages and practices of an "advertising age." See Charles Dickens, 'Trading in Death' (from: Household Words, 27 November 1852), in *Gone Astray and Other Papers from Household Words*, ed. by Michael Slater (London, 1998), 95-105; Charles Dickens, 'Betting Shops' (from: Household Words, 26 June 1852), in *Charles Dickens: Selected Journalism, 1850-1870*, ed. by David Pascoe (London, 1997), 546-52.

[3] See *The Times*, 29 August 1885, 9 ("We live in an advertising age"); *The Times*, 2 December 1889, 9 ("We live in an age of advertisement").

[4] See Phineas T. Barnum, *The Life of T. Barnum, Written by Himself* (New York, 1855); Phineas T. Barnum, *Struggles and Triumphs: Forty Years' Recollections of T. Barnum* (Buffalo, 1873); Andrea Dennet, *Weird and Wonderful: the Dime Museum in America* (New York, 1997); Neil Harris, *Humbug: the Art of T. Barnum* (Chicago, 1981); Irving Wallace, *The Fabulous Showman: the Life and Times of T. Barnum* (New York, 1959).

[5] Quoted in 'Barnum for President', *Punch*, 29 (1855), 89.

[6] 'Barnum's Elephant', *Punch*, 28 (1855), 199; 'Barnum's best plan', *Punch*, 32 (1857), 219; 'Mr. Barnum's mission', *Punch*, 36 (1859), 27.

[7] 'The Life of T. Barnum', *The Times*, 5 January 1855, 10.

[8] See T. Barnum, *Struggles and Triumphs. Or, Forty Years' Recollections of P T Barnum* (Harmondsworth, 1981), 147-57, 255-63; *Selected Letters of T. Barnum*, ed. by A. H. Saxon (New York, 1983), 24-32; Raymund Fitzsimons, *Barnum in London* (London, 1969), 33-67.

[9] See 'Barnum', *Fraser's Magazine for Town and Country*, 51 (1855), 213-23; 'The great American Humbug', *Tait's Edinburgh Magazine*, 22 (1855), 73-81; 'Revelations of a showman', *Blackwood's Edinburgh Magazine*, 77 (1855), 187-201.

[10] *The Times*, 31 October 1850, 4.

[11] See Peter Gurney, *Co-operative Culture and the Politics of Consumption in England, 1870-1930* (Manchester, 1996), 250-8.

[12] Hamilton Basso, *Mainstream* (London, 1944), 125.

[13] 'A Disfigurement of Dover', *The Times*, 19 October 1900, 5.

[14] See E. S. Turner, *The Shocking History of Advertising* (London, 1952), 112-3, 119-20; Terrence Nevett, *Advertising in Britain: A History* (London, 1982), 117; 'Folkstone is Perturbed over the Quaker Oats Ad', *Advertisers' Review*, 9 July 1900, 2; 'Abuses of Public Advertising', *Advertisers' Review*, 19 November 1900, 3-4; 'Defacing the Cliffs', *Dover Telegraph*, 18 July 1900, 6.

[15] 'Let 'em All Come!', *Daily Mail*, 2 November 1901, 8; 'A British Factory Employing over 2,000 British People', *Daily Mail*, 9 November 1901, 8; 'Don't Be "Gulled"', *Daily Mail*, 30 November 1901, 8.

[16] 'The Real British Bulldog, Not the Sham!', *Daily Mail*, 19 December 1901, 8.
[17] 'Who said "Gull"?', *Daily Mail*, 30 November 1901, 8.
[18] See Player's Cigarettes calling on "British Patriots" to smoke Players, *Daily Mail*, 1 November 1901, 8. See also Matthew Hilton, *Smoking in British Popular Culture, 1800-2000: Perfect Pleasures* (Manchester, 2000), 86-9; Howard Cox, *The Global Cigarette: Origins and Evolution of British American Tobacco, 1880-1945* (Oxford, 2000); Maurice Corina, *Trust in Tobacco: the Anglo-American Struggle for Power* (London, 1975); B. W. E. Alford, *W. D. & H. O. Wills and the Development of the UK Tobacco Industry* (London, 1973).
[19] See the full-page advertisement in *The Times*, 2 December 1901, 12.
[20] F. A. McKenzie, *The American Invaders* (London, 1902), ix-x; 1f., 8. See also Frank Mort, 'Paths to Mass Consumption', in *Commercial Cultures: Economies, Practices, Spaces*, ed. by Peter Jackson, Michelle Lowe, Daniel Miller and Frank Mort (Oxford, 2000), 7-13.
[21] See David Cannadine, *The Decline and Fall of the British Aristocracy* (London, 1992), 331.
[22] See 'Advertisement', in *Encyclopaedia Britannica, Eleventh Edition* (Cambridge, 1910), Vol. 1, 235-41; article 'Newspaper', ibid, Vol. 19, 544-81.
[23] For background see Daniel Snowman, *Kissing Cousins: An Interpretation of British and American Culture, 1945-1975* (London, 1977); Eric Kaufmann, *The Rise and Fall of Anglo-America* (Cambridge, Mass., 2004); *Twentieth-Century Anglo-American Relations*, ed. by Jonathan Hollowell (Basingstoke, 2001); Christopher Hitchens, *Blood, Class and Nostalgia: Anglo-American Ironies* (London, 1990).
[24] Mark Twain, *A Connecticut Yankee in King Arthur's Court* (London, 1889).
[25] H. G. Wells, *Tono Bungay* (London, 1953 [1908]), 136, 208. See also T. A. B. Corley, 'Interactions between the British and American patent medicine industries, 1708-1914', *Business and Economic History*, 16 (1987), 111-28.
[26] George Orwell, *Keep the Aspidistra Flying* (London, 1936), 65-6.
[27] *Advertising World* (July 1902), 88-9, 113; Terence Nevett, 'American influences on British advertising before 1920', in *Historical Perspectives in Marketing: Essays in Honor of Stanley C. Hollander*, ed. by Terence Nevett and Ronald Fullerton (Lexington, Mass., 1988), 223-40.
[28] For the history of American agencies in Britain see Douglas West, 'The Growth and Development of the Advertising Industry within the United Kingdom, 1920-1970' (PhD Diss., University of Leeds, 1984), 189ff.; idem, 'From T-Square to T-Plan: the London Office of the J. Walter Thompson Advertising Agency, 1919-1970', *Business History*, 29 (1987), 199-217; idem, 'Multinational Competition in the British Advertising Agency Business, 1936-87', *Business History Review*, 62 (1988), 467-501; Nevett, *Advertising in Britain,* 99-103; Ian Keil and Elizabeth Hennessy, 'Dorland: its Origin and Growth until 1939', *European Journal of Marketing*, 22 (1988), 49-67.
[29] Otis Pease, *The Responsibilities of American Advertising: Private Control and Public Influence, 1920-1960* (New Haven, 1958), 8f., 72.
[30] See 'The British Delegates at Boston. The Convention of the Associated Advertising Clubs of America', *Advertising World* (August 1911), 154.

[31] See 'Conventions in Advertising', *Advertising World* (September 1911), 249f.
[32] See Pamela Walker Laird, *Advertising Progress: American Business and the Rise of Consumer Marketing* (Baltimore, 1998), 304-28; Daniel Pope, *The Making of Modern Advertising* (New York, 1983), 202-12.
[33] 'World Advertising Convention: 4000 Delegates for Wembley', *The Times*, 21 June 1924, 9; 'American Advertising Men', *The Times*, 25 June 1924, 16.
[34] For the Wembley Empire Exhibitions in 1924/25 see John McKenzie, *Propaganda and Empire: the Manipulation of British Public Opinion, 1880-1960* (Manchester, 1984), 96-120.
[35] See 'The Power of Publicity: Advertising and Modern Life', *The Times*, 12 July 1924, 13; 'The Advertising Convention: War Denounced. An Outburst of Enthusiasm', *The Times*, 18 July 1924, 5.
[36] 'King George's Message', *Advertising World* (July 1924), 391; 'The Greatest Convention', *Advertising World* (August 1924), 403-6.
[37] See John C. Oswald, *The Advertising Year Book for 1924* (London, 1925), 88-92, 371-84.
[38] Ibid., 350-70.
[39] Ibid., 460-5.
[40] See *The Bill Poster*, 15 (1900), 421f.
[41] 'On the Hoardings', *Advertising World* (January 1925), Outdoor Publicity Supplement, ii. On the "colour explosion" in 1920s American advertising see Roland Marchand, *Advertising the American Dream: Making Way for Modernity, 1920-1940* (Berkeley: University of California Press, 1985), 120ff.
[42] See for example 'Enter the New Kid', *Advertiser's Weekly* (21 December 1923), 538.
[43] See 'What Simplification Means', *Advertising World* (October 1925), 25; 'New Information and Research Agency to Use USA Methods: Investigators Will Go Right into Housewives' Homes', *Advertiser's Weekly* (29 December 1938), 449.
[44] 'Blackpool May Provide British Advertising with some Shocks: W. Buchanan-Taylor Says UK Is Years behind USA', *Advertisers' Weekly* (4 May 1939), 125.
[45] 'Thirty Club Dinner to Mr. Wilson-Lawrenson', *Advertiser's Weekly* (26 January 1923), 108-9, 130.
[46] 'Sir Charles Higham Prophesies', *Advertiser's Weekly* (28 July 1932), 126. For Higham's biography see Terrence Nevett, 'Charles Frederick Higham', in *Dictionary of Business Biography: A Biographical Dictionary of Business Leaders Active in Britain in the Period 1860 – 1980*, ed. by David Jeremy (London, 1985), III, 220-2; Obituary 'Sir Charles Higham: Development of Advertising', *The Times*, 27 December 1938, 10.
[47] See again Oswald, *Advertising Year Book*, 107f.
[48] Frith Milward, 'Those Awful Supersalesmen', *Advertising World* (July 1927), 428.
[49] 'Symposium', *Advertising World* (December 1925), 259-62.
[50] 'American Advertising Has Got Little on Us', *Advertiser's Weekly* (July 27 1939), 100; Don Francisco, 'Modern American Advertising: Difference of Form but not of Function', *The Times*, 8 June 1939, xxx (United States Supplement).
[51] In 1969, the then largest British advertising agency London Press Exchange was taken over by Leo Burnett Chicago. See Giles Smith, 'LPE to Merge with US Agency', *The*

Times, 12 May 1969, 21; 'LPE Link: Good Client Response', *Advertiser's Weekly* (16 May 1969), 4.

52 West, 'Multinational Competition'; Turner, *The shocking history*, 146-52; Nevett, *Advertising*, 195; 'Why Madison Avenue moved in', *The Times*, 26 June 1969, 25.

53 Francis Williams, *The American Invasion: On American Penetration of British Economy* (London, 1962); John Pearson and Graham Turner, *The Persuasion Industry* (London, 1965); Nick Tiratsoo, 'Limits of Americanisation: the United States Productivity Gospel in Britain', in *Moments of modernity: reconstructing Britain, 1945-1964*, ed. by Becky Conekin, Frank Mort and Chris Waters (London, 1999), 96-113, 262-64; Dominic Strinati, 'The Taste of America: Americanization and Popular Culture in Britain', in *Come on Down? Popular Media Culture in Post-war Britain*, ed. by Dominic Strinati and Stephen Wagg (London, 1992), 46-81; Lawrence Black, *The Political Culture of the Left in Affluent Britain, 1951-64: Old Labour, New Britain?* (Basingstoke, 2003), 104-18; Stefan Schwarzkopf, 'They Do it with Mirrors: Advertising and British Cold War Consumer Politics', *Contemporary British History*, 19 (June 2005), 133-50.

54 Frank Mort, 'Paths to Mass Consumption: Britain and the USA since 1945', in *Buy this Book: Studies in Advertising and Consumption*, ed. by Mica Nava et al (London, 1997), 15-33.

55 'Anglo-U.S. Advertising Committee: Advice for Exporters', *The Times*, 30 September 1949; 6.

56 'The English for Madison Avenue is St. Martin's Lane', *The Times*, 28 November 1960, 11.

57 See *The LPE Reporter*, 3 (Spring 1963), 16 (copy at History of Advertising Trust Archive, LPE 3/4/1); *In and Out the Lane – IAOTL*, 3 (1962), 30; 'New trends in advertising industry', *The Times*, 27 February 1962, 16.

58 *IAOTL - In and Out the Lane*, 1 (September 1959), 14-7 (HAT LPE 3/2/1).

59 On notions of masculinity in the creative and manufacturing industries see Michael Roper, *Masculinity and the British Organization Man since 1945* (Oxford, 1994), 23-33, 132-9; Sean Nixon, *Advertising Cultures: Gender, Commerce, Creativity* (London: Sage, 2003), 95-115; Sean Nixon, 'Advertising Executives as Modern Men: Masculinity and the UK Advertising Industry in the 1980s', in *Buy this book*, ed. by Nava and others, 103-119; Jeremy Tunstall, *The Advertising Man in London Advertising Agencies* (London, 1964), 65-70, 193-8.

60 'Let's Do the Madison', *IAOTL - In and Out the Lane*, 9 (1964), 18-21.

61 *Markets and People* had started in 1946 as the *BETRO Review*, the journal of the British Export Trade Research Organisation (BETRO).

62 'Thinking out Loud', *Markets and People* (March 1951), 10-11.

63 Telephone interview with Joan Bakewell, 13.7.2004; and Joan Bakewell, *The Centre of the Bed* (London, 2003), 127-9.

64 The series was called *Dollar Sales*. Copies are at the British Library and in the papers of the Dollar Exports Board at the Modern Records Centre, University of Warwick (GB 0152 MSS.200/DEC). The Dollar Exports Board was set up at in 1949 at the instigation of the Board of Trade with the active participation of the Federation of British Industries,

the National Union of Manufacturers, the TUC and the City Advisory Group. It encouraged and assisted manufacturers to export to the "hard" currency areas of Canada, the USA and the Caribbean. The DEB became the Dollar Exports Council in December 1951. In 1959 it was reconstituted as the Western Hemisphere Exports Council, becoming a constituent of the British National Export Council in 1965. See Alan Crookham and others, *The Confederation of British Industry and Predecessor Archives* (Coventry: University of Warwick, 1997); Richard Toye, *The Labour Party and the Planned Economy* (Rochester, 2003), 158; Richard Clarke, *Anglo-American Economic Collaboration in War and Peace* (Oxford, 1982), 74; Alec Cairncross, *The British Economy since 1945* (Oxford, 1995), 52-7; Alec Cairncross, *Years of Recovery: British Economic Policy, 1945-51* (London, 1985), 73-5.

[65] Dollar Exports Board, *Dollar Sales: Advertising and Sales Promotion. A Practical Guide to the Advertising and Merchandising of British Goods in the USA and Canada* (London, 1950), 10f.

[66] British Export Trade Research Organisation, *Distributing Goods in the USA: Report on Distribution Margins and Practices, Illustrated by Case Studies* (London, 1952); 'American Market Supplement' of *Advertiser's Weekly*, (23 March 1950); 'American Market Supplement, No. 2' of *Advertiser's Weekly* (28 December 1950). The articles in the two latter publications were all written by American specialists on the subjects of market research, packaging and product design, consumer psychology, supermarket retailing, trade exhibitions, campaign planning etc.

[67] Sloan Wilson, *The Man in the Gray Flannel Suit* (New York, 1955); David Castronovo, *Beyond the Gray Flannel Suit: Books from the 1950s that Made American Culture* (New York, 2004); William Whyte, *The Organization Man* (New York, 1956), 63-78, 205-30, 312-29; Vance Packard, *The Hidden Persuaders* (Harmondsworth, 1960 [1957]), 9 warns his British readers that while his book was mainly based on research conducted in the USA, they had "little ground for complacency" as Britain had its "home-grown manipulators."

CHAPTER NINE

"THE THOUSAND GLASSY EYES": BRITISHNESS AND AMERICAN CULTURE IN TRAVEL NARRATIVES AND CULTURAL CRITICISM

BROOK MILLER

We are often quite unaware of the accommodations forced upon people the world over by America's emergence as an economic, military, and cultural "superpower" in the twentieth century. For the superpower of the nineteenth century, Great Britain, America's rapid rise engendered a wide range of anxious responses that sought to make sense of the United States in terms of its relation to Britain. My argument is that related changes occur in British self-representation during this period, and that texts that feature Anglo-American relationships provide fertile material for documenting the rise of contemporary British attitudes toward nationality. That is, looking at Americans in British texts oddly provides a key for understanding how the British perceive themselves, even today. In writing about America, British writers intervened in the discourse of what Britishness was and might be.

The distinction between nationalism and nationality is a key component of these interventions. By nationalism I mean any act of communication which attempts to legitimate an actual or desired system of government over a people. Some nationalisms support the powers that be—as American nationalism—while other nationalisms support the possibility of a new power, as did confederate nationalism before the beginning of the American Civil War. Nationalism, then, is programmatic, seeking a particular end. On the other hand, nationality is more elusive. It consists first of the relation of an individual to her or his nation. This relation consists of legal obligations, but it is typically thought of in terms of identity. One's language, one's ideals, and one's ethnicity can all contribute to a sense of nationality, as do a range of cultural stories.

The problem with this kind of identity, of course, is that it is personal and subjective. Your neighbour's accent, ideals, or appreciation of national myths may be quite different from your own experience. When you project these differences across an entire nation, you find a hectic patchwork rather than an epic tapestry. For this reason, nationality is often *nostalgic*. That is, it explains the disunities of the present moment—your differences from your neighbor—away by viewing the past as a period of unity disturbed by the forces of modern life. As a result, literature that focuses upon conveying a sense of national identity often hearkens to an earlier moment in history, and it is often structured in terms of threats to the nation by alien forces. The result, then, is a divided representation of the national self—the nation as it supposedly was and the nation as it exists, straining to return to a more "organic" state. Homi Bhabha puts the matter thus: the nation is divided by "the tension signifying the people as an *a priori* historical presence, a pedagogical object; and the people constructed in the performance of narrative, its enunciatory 'present.'"[1]

The lingering problem with nationality, experienced in this way, is consciousness of the split between the two types of identification. In literature promoting an ameliorated sense of nationality—and in the modern era, of course, "literary" fiction often features a modern literary sensibility that represents the best model for the national self—writers explore ways to heal or obscure this split. One common way of doing this is to characterize the "performative" contemporary subject as besieged simultaneously by outside forces, especially outside economic forces, and degenerated or corrupted elements of her/his own society. And that's where the United States plays a crucial role for late nineteenth- and early twentieth-century writers.

The character of the United States has been a subject of intense scrutiny in British writing literally since the day the American Revolutionary War ended. At the end of the nineteenth century, however, two notable phenomena occur. First, the number of published travel narratives about the U.S. more than doubles; second, America and Americans become staples of British fiction. Whether writing about the realities of domestic social life or fantasizing about the far reaches of the Empire, British writers brought America and Americans into the fabric of their fictions even when their subject had little to do with the U.S.

To account for these phenomena, we may refer first to changes within British society, and second to the changing nature of Anglo-American relations. In a well known collection of essays entitled *The Invention of Tradition*, Eric Hobsbawm and Terence Ranger argue that the period between 1870 and 1914 witnessed the abundant growth of new traditions in British and, more widely, European society. These new rituals, emblems, and institutions promoted a national cultural ideal accessible to legions of newly literate citizens and helped reconcile antagonistic social groups by offering images of aristocratic privilege as the shared weal of the public. Meanwhile, Anglo-American relations shifted dramatically as well. Residual tensions over the Civil War, Britain's increased diplomatic isolation with the international politics of imperialism, a border dispute in South America, debates about the gold standard, Irish politics, America's growing debt to Britain, American refusal to comply with international copyright standards, and Britain's relative decline as a sea-power and industrial producer all impacted exchanges between the two nations. Indeed, several British politicians and commentators suggested a formalized political bond between the nations, partly as a way of sustaining British influence. The actual result was significant, if less dramatic: at the turn of the century, the term "The Special Relationship" was coined to designate a prioritization of partnership with the United States over other diplomatic obligations. This all at the moment when Britain was putatively at the height of its imperial power.

These political developments were accompanied by social developments as well and the period witnessed a proliferation of race discourse about Anglo-Saxon superiority and the role common "Teutonic origins" played in making the two nations beacons of liberty and progress. This mutual affirmation led to rather surprising results: at the close of the nineteenth century, for example, municipal governments held American Independence Day celebrations in London and many of Britain's smaller cities!

During this dynamic period, British travel narratives shifted from comparative analysis to what Daniel Rodgers calls an "aesthetic framing whose keys…were culture, custom, and time."[2] In the period when American industrial and military growth began to impact British dominance, British readers took comfort in assessments such as the following from the influential historian and diplomat James Bryce: "the gray monotony of American life [made]…too few local variants, and there was less difference between a Mississippian and an Oregonian than between a Yorkshireman and a Cornishman."[3] The critique of homogeneity, which replaced scathing assaults on Americans' lack of manners, involved a paradox: on the one hand, they critiqued a nation excessively driven by nationalist doctrine; on the other, they aimed to obliquely evoke a sense of national difference by focusing upon the distinctions and nuances within British, even English, subcultures. The inconsistent impulse betrays anxiety that the forces of modernization threatened the stability of culture and tradition. Indeed, the historical realities support such a claim. Rodgers argues, "by the last quarter of the nineteenth century there were…uprooted peasants" in massive numbers in both the U.S. and Europe: "in England by the century's end, scarcely more than a quarter of the people lived in the county in which they had been born."[4]

In the context of this internal flux and external competition from the U.S. and Germany, British texts about America functioned refashioned pragmatist images of John Bull into celebrations of British cultural distinction. The signature of these texts is that the narrator approaches his or her analysis evaluating America through a position of knowing mastery towards which she or he seeks to lead the reader. The refinement of personal faculties of judgment and taste provide a non-economic, non-sectarian, and indeed national locus of value expressed through the critique of an other. In claiming identification with the British reader in terms of personal faculties of judgment and taste, the narrator imaginatively reconciles disparate political and social positions. Indeed, writers from widely different social and political perspectives deploy this version of British identity during the period under consideration. In the remainder of the essay I will consider the writings of three very different figures whose work illustrates these dynamics.

In Matthew Arnold's "A Word on America" (1885) and "Civilization in the United States" (1888), Beatrice Webb's *American Diary* (1898), and Henry James' *The American Scene (1907)*, the critique of America provides this uniting function. The content of the critiques, moreover, share a central paradox—on the one hand, the authors relate the self-conscious anxieties of Americans to the textual, tautological nature of American national legitimacy[5]; on the other hand, homogeneity is related to the basically economic nature of American cultural aspirations. That is, these critiques simultaneously invoke a materialist view of U.S. culture and a critique of its idealist ideology. The central assumption of these critiques is that America's flaws stem from the void of an absent, or erased, cultural tradition. In asserting a materialist perspective upon American culture, these authors implicitly invoke the redemptive authority of English culture. Their quasi-Marxian cultural analysis is framed by a cultural-nationalist view of their own identity.

This analysis supplements one of Fredric Jameson's central contributions to revision of Edward Said's analysis of orientalism: in the late Victorian and modernist periods, the discourse of alterity central to British nationalism focused upon imperial competitors rather than the colonial subject. As Jameson puts it in "Modernism and Imperialism," "during this period 'imperialism' designates, not the relationship of metropolis to colony, but rather the rivalry of the various imperial and metropolitan nation-states among themselves… The proto-typical paradigm of the Other in the late nineteenth century… is the imperial nation-state."[6] In the texts considered in this essay, thematic unities in the representation of Americans reflect a common fund of assumptions shared between narrators and implicitly attributed to British readers. While each narrator investigates American culture and performs what Žižek calls "symptomatic critique" of how culture reflects material and historical conditions, this essay analyzes the implicit narrative persona of each text to perform a culturalist, psychoanalytic reading Žižek describes as "extracting the kernel of enjoyment… articulating the way in which—beyond the field of meaning but at the same time internal to it—an ideology implies, manipulates, produces a pre-ideological enjoyment structured in fantasy."[7] In the texts considered nationality functions as a *point de capiton*, or quilting point, in the construction of American difference. For Arnold, Webb, and James, the "enjoyment" of their writing involves the instruction of and identification with an implied middle-class British reader. In doing so, these writers articulated early critiques of mass culture that share remarkable similarities with contemporary cultural criticism.

As a cultural critic, Matthew Arnold is best remembered for *Culture and Anarchy*, in which he warned that England's rigid class system, and the insufficiencies of each class, threatened to dissolve the nation into anarchy. He prescribed a renewed engagement with "the best that has been thought or spoken", or "culture." In *Culture and Anarchy* Arnold is also highly critical of the United States, describing Americans as provincial and warning English readers that such a fate may await them as well. He promotes culture—embodied, particularly, in literacy and social criticism grounded in the traditions of English literature—as an ameliorative force in an era increasingly dominated by mass politics. Against accusations of the passivity of the project he's outlining, he describes men of culture as "firmly bent on trying to find…a firmer and sounder basis for future practice…and believing this search…to be…of yet more vital and pressing importance than practice itself…we may…do more,…we poor disparaged followers of culture, to make…the frame of society in which we live, solid and seaworthy."[8] This pedagogical function for culture, and criticism especially, makes Arnold a proponent of literature, art, and other elements of 'civilization' as essential to reshaping the nation and avoiding political catastrophe. For Arnold culture was attractive because it offered "disinterestedness" (rather than, for him, class interest) as a basis for non-political writing. Arnold perhaps differed from most contemporary critics in his belief in the ameliorative possibilities of "right reason," but he participated in the same dilemmas academics face when they are asked to defend the serious study of cultural forms.

After his rather scathing commentary on the U.S. in *Culture and Anarchy*, in the 1880's Arnold penned three essays focused specifically upon the United States, "A Word About America," "A Word More About America," and finally "Civilization in the United States." In these texts he moves from the criticism that Americans are provincial to an emphasis, finally, upon the want of "beauty" and "distinction" in American culture. As his prose undergoes this transformation Arnold continues to insist that England needs reform, and that his examination of America fulfills the purpose of helping him shape his reform programme.

The most striking feature of Arnold's three pieces is that he consistently critiques England and writes in response to other writers, in particular George Bancroft, whom he describes as America's national historian, Sir Lepel Griffin, an Indian colonial administrator, and a variety of editorialists, including James Lowell. In many respects Arnold's comments in these essays represent a retreat from the strong positions regarding American provincialism he offered in *Culture and Anarchy*. In "A Word About America," for example, he continually asserts that America and England, while similar, are characterized by distinctive class compositions; America lacks a class of gentlemen and is dominated by the Philistinic middle class. He begins the essay by rebutting an American pressman from the *Atlantic Monthly* who has excoriated Arnold for attacking the U.S. Arnold quotes the article, in which the reporter writes that "the hideousness and vulgarity of American manners are undeniabl[e]," and that "redemption is only to be expected by the work of a few enthusiastic individuals, conscious of cultivated tastes and generous desires."[9] Here Arnold seems to perform the good taste he claims derives from high literary standards by letting crude, direct criticisms of America be articulated by the pressman. That is, he critiques by withdrawing from the debate.

At the same time, for Arnold American anxiety is expressed in a lack of control over text and textuality. Arnold points out the pressman's desire to make assertions without carefully studying the text. Such impetuous behavior reveals, in Arnold's calculus, what happens when a society is cut off from the mainstream of culture. This rhetorical strategy of patiently responding to misguided criticisms of the United States and of his own writings is important for his cultural politics. By exposing the mistaken assumptions and misreadings of others in an exasperatingly coy form of politeness, Arnold foregrounds his aesthetic judgment and good manners as emblems of the civilization he finds wanting in the United States.

Arnold asserts that he "has long accustomed [himself] to regard the people of the United States as just the same people with ourselves, as simply 'the English on the other side of the Atlantic'."[10] In his view, America is different only in terms of its class composition—while England has an aristocracy and working classes, America seems to consist entirely of a vast middle class, the Philistines. American Philistines are "the great bulk of the nation, a livelier sort of Philistines than ours, and with the pressure and the false ideal of the Barbarians taken away, but left all the more to himself, and to have his full swing."[11] Here aristocratic "false ideals" do not hold sway, but Philistine hegemony results in an excessive valuation of the "powers of industry and conduct." For Arnold, "the building up of human life, as men are now beginning to see, there are needed ... the power ... of intellect and knowledge, the power of beauty, the power of social life and manners. And that type of life which our middle class in England are in possession is one by which neither the claims of intellect and knowledge are satisfied, nor the claim of beauty, nor the claims of social life and manners."[12] This negative assessment of American and English philistinism comes in the larger context of a critique of class divisions, and on this subject Arnold makes a key distinction between the two countries.

In "A Word More about America," Arnold argues that a rigid class structure "hampers and falsens" individual thinking, and he recommends establishing social institutions on the basis of ideals rather than "by what one can oneself gain from it."[13] He praises the workings of American institutions insofar as they are not riven by ideological divides based upon class difference—that is, insofar as they reflect the perceived homogeneity of the American people.

While homogeneity is conducive to good institutions, it is also a key target in his final writing on Americans, "Civilization in the United States." Arnold's ambivalence towards homogeneity reflects a moderation of his position in *Culture and Anarchy*. Indeed, it reveals that as Arnold became clearer about the role of Culture he began to view it in aesthetic rather than socio-political terms. Insofar as this shift mirrors a generally growing unease about American power on the global stage, it reflects a movement towards the representation of America and England in symbiotic relations not for *realpolitik* purposes but because each nation's character could fill the lack of its opposite in a mutually useful relation. Simply put, this model imagined the merger of American energy with British refinement.

In this final essay, Arnold isolates the key notes of civilization, locating "the great sources of the *interesting* [in] distinction and beauty: that which is elevated, and that which is beautiful." To Arnold, America has "little to nourish or delight the sense of beauty," particularly in urban areas. Conversely, Arnold believes Americans are "restless, eager to better themselves and to make fortunes" and thus "the inhabitant does not strike his roots lovingly down into the soil, as in England." English difference thus stems partly from an identification with place, while American rootlessness and materialism are mutually reinforcing. Moreover, Arnold returns to this claim that America is pervaded by Philistinism because that Americans "came originally, for the most part, from that great class in English society amongst whom the sense for conduct and business is much more strongly developed than the sense for beauty."[14]

To illustrate his claims, Arnold points to Abraham Lincoln. Lincoln becomes an emblem for the "typical American…shrewd, sagacious, humorous, honest, courageous, [and] firm…but he has not distinction."[15] Lincoln's stolid nature exemplifies the serious "serious philistine." Distinction is an ambiguous quality, though it is in part the "discipline of awe and respect" developed through contact with elevated cultural objects.

Arnold lambastes the American press as a prime cause of this want of distinction, and his rhetorical method of making arguments through response to inflated claims in the press underscores the degree to which his essays perform the results of contact with the national culture he has described. He reiterates his critique of England while making his concluding remarks about America: "I once said that in England the born lover of ideas and of light could not but feel that the sky over his head is of brass and iron. And so I say that, in America, he who craves for the *interesting* in civilization, he who requires from what surrounds him satisfaction for his sense of beauty, his sense for elevation, will feel the sky over his head to be of brass and iron."[16] How are we to take the distinction Arnold makes here between the diagnosis of England in *Culture and Anarchy* and the diagnosis of the U.S.? The difference resides in subtleties of phrasing: whereas he considers the alienation of English *lovers* of ideas and light, he describes Americans as "craving" and "requiring," in other words as "lacking." He suggests that whereas England's lack is the English's, America's lack is America's. That is, whereas the modern Englishman has strayed from the love of culture, the American is imprisoned in a dessicated cultural landscape which permits no civilization. English culture requires a *return*, while American culture remains an eternal oxymoron.

Later critics, including Beatrice Webb and Henry James, were influenced by Arnold's widely read articles. They too fix upon articulating British culture as a nostalgic ideal to which the subject might return but against which American materialism is positioned. Where they differ, it seems, is at three points: in the treatment of homogeneity in the guise of the crowded urban experience, in confronting America's growing nationalistic fervor and commercialism, and in shaping their performance of British identity in terms of different political and aesthetic agendas.

Webb wrote her 1898 *American Diary* while visiting the United States to study the workings of local government. In it, she joins a more conservative group of travel writers who claim that America's idealization of liberty is a form of bad faith. Her diary records her impressions of Americans and their society, often linking generalizations about their character to the representative and legislative styles they've chosen. Her lasting impression of urban life in America is of "Noise, noise, nothing but noise…your senses are disturbed, your ears are deafened, your eyes are wearied…your nerves are shaken and rattled…."[17] The intensity of urban experience, in its tendency towards sensory overload, leads Webb to identify a variety of paradoxes related to America's simultaneous restlessness and stagnation. Americans, forged in the din of a homogenous, electric crowd are at once intelligent and unintellectual, from diverse origins and "wearisome[ly] homogenous, and possessed by a "capacity for prompt and unhesitating action" which is "the other side of their gravest national defect—impatience."[18]

Webb's survey of the United States includes an acute critique of American nationalism. In San Francisco Webb attends a July 4th celebration that includes a recitation of the Declaration of Independence. She observes that the celebration promotes American exceptionalism: "the note of the whole thing was the unique character of American Institutions…the Americans being the chosen people who had, by their own greatness…discovered freedom, and who were now to carry it to other races." She marvels at the fanatical response of the American crowd to the reading, "as if this declaration had been made yesterday and sealed by the blood of the present generation." Webb emphasizes Americans' fetishistic worship of the nation's founding documents. Behind this she detects "how anxious [patriotic sentiment seems] to assure itself of the nation's glory and how swaddled up in abstract propositions which have long since lost their meaning."[19] Webb's detection of anxieties beneath patriotic fervor makes her an antecedent of current critical work on nationalism. Moreover, her attention to nationalism as a socio-political force, and her attempt to historicize it as a mass-cultural phenomenon place her within this genealogy. However, the conclusions she draws about the American national character reveal that these analyses are bounded by an interest in distinguishing English from American capacities for culture in a manner which is neither "materialist" nor "idealist," but based in ideas about racial types.

While her politics differ dramatically from her more conservative contemporaries, Webb makes similar claims about the American temperament. She notes "the absence of distinguished talent and of variety of type of the American race." However, this homogeneity is attributed to ideological causes: she supports "the hypothesis that there is something in the way of life and the mental environment of the U.S.A. that hinders or damps down the emergence of intellectual or artistic distinction."[20] Like her contemporaries, Webb associates the U.S. with a want of culture, in this case meaning a lack of artistic achievement. Yet she holds the idealistic perspective that American ideology is the prime cause of homogeneity. She bases this hypothesis upon two fallacies of American thought—that is, unlike critics who view American ideals as pervaded by bad faith in their application (and as a result, often, of their hybrid racial nature), for Webb the fault lies within the ideals themselves. She attributes this lack to two fallacies:

> that "all men are born free and equal;" with its derivative that one man is as good as another and equally fitted to deliver judgment on every conceivable question… It is… Protestantism, [the] assumption that there is no such thing as the expert, and that all men are equally good judges in all questions, that is one of the fallacies that eats away the roots of any American genius by blinding his fellow citizens to his peculiar talent or attainments… The Englishman… has inherited sufficient experience of affairs to reject the rotten metaphysicals of Protestantism and to perceive that the whole modern theory of division of labour leads straight to increased specialization of particular faculties in particular

> people, and to the careful selection of experts for the finer work of a complicated civilization.[21]

In her comparison of the Englishman and the American Webb thus valorizes specialization as a marker of an advanced culture. If it seems peculiar that Webb thereby insinuates that intellectual and artistic distinction are somehow fostered by a society with an advanced division of labor, we might attribute this to her Fabianism. Regardless, for Webb an advanced division of labor contributes to a strong sense of the state. Her second fallacy of American ideology reflects this position:

> The second assumption is perhaps less consciously held, but is more universally acted upon. It is the old fallacy of the classic economists that each man will best serve the interests of the whole community by pursuing his own gain. It is interesting to note that this axiom was invented by English thinkers; but it has never been fully accepted by the English people.[22]

Webb associates American ideals with a naïve individualism. Her view aligns her with the Arnoldian vision of a culture of experts. She depicts America as a society in which the anarchy of a lack of proper social distinctions and a lack of specialization defeat any promise of cultural development. America is disturbingly homogenous: "all professions, all occupations, resolve themselves into a race for money. However diverse may be the origin and physical environment of a people, if they have but one motive there can be but one faculty. Hence the all-pervading and all-devouring 'executive' capacity of the American people." For Webb, "the American has less capacity for deliberately organized life than the Englishman or the German" because he lacks the qualities of long-sightedness, careful discrimination and judgment, and rationality.[23] The editor of Webb's diary, David Shannon, points out that her strongest conclusions, such as these, stem from a cultural myopia: "So British was she that she was quite incapable of understanding non-Britons. Thus, although she tended to judge American politicians by their intellectual powers or lack of them, she quite consistently judged American intellectuals by nonintellectual standards, primarily by their appearance and manners as compared to their British counterparts…."[24]

Webb focuses, like so many of her counterparts, upon American businessmen as representative figures of what is wrong in American society. Businessmen lack standards of excellence because they lack culture. She notes the "extreme conventionality" of their ideas, and asserts that although the prevailing ideology is individualistic, there is in practice great "contempt for the vested interests… of the individual citizen. Private enterprise is permitted to trample on the individual."[25] Webb alludes to a common complaint within American society in the wake of the struggles against monopoly capital in the

1890s: corporations are vested with the rights of individual citizens, and thus the citizen finds her/his own rights circumscribed within contractual relations with an unimpeachable, unassailable corporate might. While such an observation reflects her socialist leanings, Webb's Arnoldian affinities are apparent in her consistent assignations of character to an emergent social type: the "intellectual traditions of the ordinary American business man are a naïve individualism tempered by an opportunist consideration for any new forces that may appear… he is… an ideal philistine." Webb describes this type as a "rule of thumb man, destitute of all culture."[26] Webb's connection of "rule of thumb" pragmatism associated with earlier versions of John Bull and Arnoldian philistinism in the figure of the American represents, if not a direct displacement of anxieties about British massification, at least a rather direct attempt to associate Britishness with an anti-philistine character.

Like Webb, in *The American Scene* Henry James focuses upon urban masses as a key to understanding American identity. As he rides a train into New York City, James takes pleasure in observing its vitality, and indeed sees a superficial harmony between nature and science in "the extent, the ease, the energy, the quantity and number."[27] Once James is walking through the city, however, this harmony is badly disfigured by the stark, overwhelming presence of skyscrapers and the pulsing masses along Wall Street. James views the skyscrapers as America's repugnant "contribution" to the traditions of European architecture. They become "the most piercing notes in that concert of the expensively provisional into which your supreme sense of New York resolves itself."[28] This provisionality reflects America's lack of history and, further, the lack of "credible possibility of time for history" caused, in his view, by the predominance of commercialism. He sees a pervasive "ugliness" in both rural and urban landscapes caused by what James calls "the complete abolition of forms."[29] In this paradox—America is "history-less" but forms have been "abolished" rather than simply absent—we see the Jamesian emphasis upon cultivated tradition as the stimulant of new cultural production and his distaste for revolutionary movements, both political and artistic. Through a drama in which James invokes a Burkean notion of the sublime, James comes to see the skyscrapers as a "conspiracy against the very idea of the ancient graces."[30] As he stands before the gleaming new buildings, James begins to feel overwhelmed. He sees a linkage between pervasive commercialism and the artifices at the root of America's dramatic expansion: the skyscrapers speak with an architectural "vocabulary of thrift at any price [which] shows boundless resources, and the consciousness of that truth…of the finite, the menaced, the essentially *invented* state, twinkled ever, to my perception, in the thousand glassy eyes of these giants of the mere market." The emphasis upon menace, invention, and a pervading watchfulness sends James into a kind of sensory overload, in which his sensations "testify to the character of NY" but nevertheless "kept overflowing the possibility of poetic, of dramatic capture." James' moment of capturing the effect does come, however, when he names the sensation in terms of the discourse of the sublime: "the vast money-making structure quite horribly, quite romantically justified itself, looming through the weather with an insolent cliff-like sublimity."[31] This sublime effect comes from the merging of the skyscrapers with the weather, the way the building towers over a church, and, crucially, the experience of a pushing male crowd. In its homogeneity and density the crowd nearly overwhelms reflection in its "sounds and silences, grim, pushing, trudging silences, of the universal will to move…an appetite at any price."[32] James' victory against this crowd is artistic—he survives it and captures it in *The American Scene*. Nevertheless, the drama of the sublime foregrounds the artist's difference from Americans and indeed associates them with a horrific, unthinking appetite that is the very antithesis of James' refined, restrained ideal.

The effect of the artist's encounter with the sublime generally entail a moment of panic, in which human reason is stunned by nature's vastness. Here James deploys this logic to insinuate a sense of primitivity into the mechanization of life in America. Yet his sublime involves the horrors of a formless, incessant human aggregation. Immediately after his description of Wall Street's skyscrapers, James reports his horror at the constant flow of immigrants through NY's Ellis Island which he sees as an "act of ingurgitation on the part of our body politic and social…that never, never, never stops." The experience creates a "sense of dispossession" that haunts his sense of national belonging. The inversion implicit in this idea, that immigration robs James of his nationality, point to how closely he, along with many of his contemporaries, associated ideas about nationality with race. The result is that "free existence and good manners…are too much brought down to a bare rigour of marginal relation to the endless electric coil."[33] Thus James, like Webb before him, employs electrical metaphors to convey his sense of America's energy and subordination to commercial interests.

The converse of James' association of Americans with a monstrous body of unceasing appetite is his portrait of the commercial equivalent of the cultural ideal. In contrast to the messy, voracious muddle of the streets of New York, in Florida James finds quiet and a semblance of good manners in the opulence and sense of enclosure of an expensive seaside hotel. At first the Poinciara Hotel strikes James as a "supreme illustration of manners of a social 'case'…[and] fresh and luminous."[34] By contrast to the horrific sublime of NY, "the sublime hotel-spirit…operates by an economy so thorough that no element of either party to the arrangement is discoverably sacrificed." An apparently generous interchange between hotelier and customer becomes emblematic for an exchange between the spirit of commerce—the "hotel-spirit"—and the spirit of national life. In this interchange "the hotel-spirit is an omniscient genius, while the character of the tributary nation is still by struggling into relatively dim self-knowledge."[35] Thus James sees the national character as a "tributary' to the spirit of material gain, as only dimly self-aware, and as ultimately "adjustable." This mutability reflects, again, the lack of history and culture to give the American a stable sense of self and, importantly, an absence of distinction and "the jealous cultivation of the common mean." This hotel-spirit creates and shapes national appetites, leading ultimately to an absence of individuality and variety, and dimness of "distinctions."[36]

James' critiques of the homogeneity and commercialization of American culture are in a sense sociological accounts of the impact of commercial values. However, it is crucial for James that this analysis not convey objective detachment, but rather an intensely personal performance of his connection to his English readers. The hotel-spirit and adjustable national life are antithetical to the spirit of his narrative persona, the "restless analyst." The analyst is an "agent of perception" who "would take [his] stand on [his] gathered impressions, since it was all for them…that [he] returned." He returns to the U.S. not so much for the place itself, but rather for the impressions which his perception shapes and gathers. He is emphatic on this point: "I would in fact go to the stake for them."[37] James' valorization of the "cultivated sense" as not merely the key to acute perceptions and good judgment but in fact as an "enrichment" of the subject matter differentiates his persona from Arnold's and Webb's position. James' man of culture, "acute as an initiated native" while ennobled by immersion in English culture, enriches reality through his refined sensibility.[38] Such a figure, whose perceptual powers in a sense make the cultural object observed, has aligned himself with an "English" readership. Indeed, he has performed his status as an British citizen by articulating one's "supreme relation" in terms of "restless analysis" rather than passive embodiment. *The American Scene*, finally, testifies to James' desired Britishness.

Thus, America's economic and political surpassing of Britain during this period is reimagined by British writers in terms of excessive, irrational, uncanny, or even monstrous growth. American culture seemed, above all, relentlessly commercial. By recognizing how the idea of national identity provides a buttress against forms of alienation, we can begin to grasp the prevalence of contradictions regarding what constitutes the British character between different texts and, importantly, within single texts. National identification, in contrast to nationalism, is at least partially, and perhaps mostly, a negative form of identification. Cultural representation tends to treat the British character as a Lacanian "by-product" during this period: a quality that cannot be directly performed or obtained, but nevertheless is constitutive. National myths of character provide a support against the very forms of alienation and disaffection that the state and corporate institutions produce. When invoked in narrative, the real alienation produced by the new imperialism was most likely to be found in the guise of other nationalities, in some cases America.

As our examples suggest, in its repressive aspects national identification embodies a form of class fantasy—a conviction that the healthy nation (often an ideal quite explicitly contrasted to the state) depends upon one's social group. In particular, national identification was important to newly educated middle classes, people whose social position was created by Britain's economic growth and imperial supremacy in the nineteenth century. Narratives of nationhood, which placed the present moment at the apex of cultural development, appealed to groups who did not fit comfortably within older class-based narratives. Yet national fantasy, like other forms of fantasy, is aspirational: often the valorized national group is not simply typified in its literary representation; rather, it is elevated. Each of our writers critiques this conflation of nation and class interests while performing a model for national culture in their narrative persona.

---

[1] Homi Bhabha, *Nations and Narration*, (New York, 1990), 298-9.
[2] Daniel T. Rodgers, *Atlantic Crossings: Social Politics in a Progressive Age* (Cambridge, Mass., 2000), 39.
[3]Bryce, James. *The American Commonwealth*. (1888. Washington D.C., 2002).
[4] Rodgers, *Atlantic Crossings,* 49
[5] I.e., "we the people" are a legitimate polity because "we" have a body of documents asserting our legitimacy. This analysis is Michael Warner's in *The Letters of the Republic: Publication and the Public Sphere in Eighteenth-Century America.* (Boston, 1990).
[6] Fredric Jameson. "Modernism and Imperialism." in *Nationalism, Colonialism, and Literature. (*Minneapolis, 1990), 48-49.
[vii] Slavoj Žižek, *The Sublime Object of Ideology.* (New York, 1989), 125.
[8] Matthew Arnold, *Culture and Anarchy*. (Michigan, 1994), 135.
[9] Matthew Arnold, "A Word on America," in *Discourses in America.* (Michigan, 1970), 1.
[10] Ibid, 2.
[11] Ibid, 7.
[12] Ibid, 10.
[13] Ibid.
[14] Matthew Arnold, "Civilization in the United States." in *Discourses in America.* (Michigan, 1970), 54-55.
[15] Ibid, 56
[16] Ibid, 58.

[17] Beatrice Webb, *American Diary*. (Madison, 1963), 136.
[18] Ibid, 154.
[19] Ibid, 138-139.
[20] Ibid, 148.
[21] Ibid, 149
[22] Ibid, 149.
[23] Ibid, 145, 150.
[24] Ibid, xii.
[25] Ibid, 13.
[26] Ibid, 13.
[27] Henry James, *The American Scene*. 1907. (New York, 1994), 153-54.
[28] Ibid, 77.
[29] Ibid, 25.
[30] Ibid, 89.
[31] Ibid, 77-82.
[32] Ibid, 83-84.
[33] Ibid, 83-84, 89.
[34] Ibid, 438.
[35] Ibid, 439.
[36] Ibid, 442.
[37] Ibid, xxv.
[38] Ibid, xxv.

CHAPTER TEN

"THE FINEST OR THE DAMNDEST COUNTRY IN THE WORLD ?": BRITISH RADICALS AND AMERICA IN THE 1930S

ROGER FAGGE[1]

Ever since the nineteenth century the United States has represented an unavoidable challenge for the British left. On the one hand they saw it as embodying the triumph of monopoly capitalism, spreading its values ever outwards across the globe. On the other, the shared language, lucrative lecture tours, friendships and positive engagement with aspects of American culture ran counter to this ideological dismissal; a situation given further impetus when elements of the British left encouraged a shift from a rhetorical opposition to capitalism in favour of an interest in managing it more effectively. This was most obvious with the "revisionists" of the 1940s and 50s, and more recently, New Labour, who represented the closest engagement with liberal, and in the case of the latter, not-so-liberal America.

Of course, prior to the large-scale expansion of US industry, British radicals had viewed America in a much more positive light, believing the American Revolution and first decades of the new republic, represented a noble republican experiment and an advance for the democratic ideal around the world. Thomas Paine's influence was particularly important in this regard, tying British and American radicalism together, and British radicals also came to admire what they saw as the democratic efficiency of government, the absence of bloodletting when compared with the French Revolution, and the good standard of living experienced by the American working man.[2] The balance shifted, however, after the Civil War, with the transformation of the US into a mass, industrial society, and what British radicals saw as the corruption of the original republican ideal of America.[3] The influence of Socialist and other radical ideas by the early twentieth-century gave further impetus to this process, providing a stronger ideological critique of industrial America. The British left's support for dissident Americans like Daniel DeLeon of the Socialist Labor Party, "Big" Bill Haywood of the Industrial Workers of the World, and Eugene V. Debs, of the Socialist Party of America, further reinforced the left's preconceptions about America. De Leon, for, example was a particular influence on British syndicalists, and Haywood was well received when he spoke in the mining valleys of south Wales.[4]

This somewhat paradoxical relationship between the British left and America has received relatively little attention from historians. Henry Pelling's clear and well argued *America and the British Left* remains one of the best studies, although that was published as long ago as 1956. Others have also written on this area, and there has been some useful work on post 1945, however there is still more to be discovered.[5] Consequently this chapter looks at the 1930s, an important period when British radicals were at their most ideological, and strongly critical of the US. However, this position came under pressure from the innovative response of Roosevelt and his New Dealers to the Great Depression, which attracted the interest of certain British radicals. In addition, a number of eminent radicals visited the US in this period, giving them an insight into a country they had previously only read about, or seen on a cinema screen. The result was, for some, a stronger engagement with America, but not necessarily a settled, or even coherent view of the influence and significance of the country.

This chapter considers both political and cultural radicals in this period, but focuses more on the latter since they produced the most useful and interesting material on the US. Cultural radicals usually travelled more widely, and were able to avoid the ideological and organisational shackles of some of their political counterparts. As we shall see, figures like H.G. Wells, J.B. Priestley and W.H. Auden in particular, offered a number of insights into the meaning and complexity of the United States, not least an acceptance of the centrality of America in the modern world, and an appreciation of the democratic tendencies within American political culture. Even here, however, America remained something of a paradox, and there was a tendency by writers to reinvent it to fit their own world-view.

By the 1930s suspicion of the power and influence of the United States was apparent in Britain. One aspect of this concerned the way the dynamism of the American economy was being felt around the world. This was apparent in Britain, where American economic, social and cultural influences were being felt. Whilst dwarfed by Americanisation after 1945, these influences (ranging from cinema, fiction, housing and roads, to broader patterns of consumerism) were significant enough to grate with native sensibilities; coinciding as they did with post-war feelings of economic and cultural insecurity in Britain. This insecurity was made worse by a sense of dented pride as Britain's star declined, whilst the US's ascended.[6]

George Orwell was one of those who captured this sense of unease. In his 1939 novel, *Coming Up for Air,* which lamented the rise of a more regimented, commercialised order, whilst at the same time reflecting on the approaching war and its totalitarian possibilities, Americanisation was symbolised by the Milk Bar which was "slick and shiny and streamlined," with "no real food at all. Just lists of stuff with American names, sort of phantom stuff that you can't taste and can hardly believe in the existence of." Orwell added "everything's streamlined nowadays, even the bullet Hitler's keeping for you."[7]

British radical misgivings about the encroaching tide of Americanisation was in part based upon a shared national presumption about the inevitability of British power (albeit of a more radical tinge in their case), however the main source for anti-Americanism came from the realisation that the US was now fulfilling many of the fears that they had voiced since the end of the nineteenth-century. The left argued that the US now clearly represented the worst manifestations of monopoly capitalism, with power in the hands of an aggressive capitalist class, bent on extending their influence across the globe. As early as 1918 the Labour Party Programme, "Labour and the New Social Order", written mainly by Sidney Webb, contained an explicit warning about the impact of this process in Britain.[8] The situation was exacerbated by the slump that hit the American economy from 1929, triggering the Great Depression, and illustrated the interconnection of the world economy, and the weakness in comparison of other industrial nations who witnessed an economic downturn of their own as a consequence.

Some on the left saw this as representing the much heralded crisis of capitalism. The icing on the anti-American cake was provided by the traumatic collapse of the second Labour Government in August 1931, after the New York Federal Reserve Bank demanded that, in return for further loans, the Government should make cuts in unemployment benefit and other expenditure in line with the proposals of the May Committee. The subsequent arguments split the Cabinet and led to the widely held belief that the New York "Bankers Ramp" had helped bring down the Government.[9]

Anti-Americanism could also coincide with pro-Sovietism on the far left, but as the 1930s unravelled the Soviet Union became a less attractive model and, at the same time America came to be seen by some on the left in a more balanced light. Franklin Roosevelt's New Deal programme was influential in this regard, representing a contrast to the "rugged individualism" of the Hoover Administration and the smug complacency that had characterised the "Jazz Age" of the 1920s. Faced with the catastrophic consequences of the Great

Depression, which by early 1933, saw a quarter of the US workforce unemployed, Roosevelt's pragmatic response, mixing old and new ideas, nevertheless represented an enhanced role for the state, both in the short term in relieving the consequences of the Great Depression, and by 1935 in the creation of a rudimentary welfare state. The latter included outcomes which were not necessarily intended by Roosevelt, as when the recognition of Trade Unions, especially through the 1935 Wagner Act, gave rise to the sit-down strikes of 1935-6 that swept through manufacturing, helped by the newly established mass, industrial union, the Congress of Industrial Organisations (CIO)—an organisation much closer to the hearts of British Trade Unionists.[10]

The New Deal's attempt at "bold experimentation" offered a distinct contrast with the conservative response of the National Government in Britain. However, many radicals, especially early on, held to the line of "no compromise" between capitalism and socialism, seeing the New Deal as corporatist solution to the problems facing capitalism that would result in fascism.[11] However, this response was complicated by two factors: Firstly, as the New Deal progressed, offering genuine reform rather than emergency measures, some came to believe that it might either offer a route to a socialist democracy, or, at the very least, a pragmatic palliative at a very dangerous moment in western history. Roosevelt's stand against fascism was important in this regard as it called into question the left's argument about the fascistic tendencies of the New Deal. One of the most influential left intellectuals was Harold Laski, whose views on the New Deal during 1930s were only consistent in their inconsistency. He blew hot and cold over the New Deal, rejecting it early on before recognizing its greater potential by 1938, writing that "I believe that the Rooseveltian experiment in America, with all its blunders and follies, will be regarded by the historian as having made a supreme contribution to the cause of freedom."[12] A similar shift occurred with others including the more rigorous Marxist John Strachey whose opposition to the New Deal had wavered by 1938.[13]

A more sympathetic approach to the New Deal was also encouraged by the strong, charismatic leadership of Roosevelt which appealed given the paucity of such figures on the British left. This was particularly true for those who met or corresponded with the President, and witnessed his charisma firsthand. Laski, for example exchanged correspondence with Roosevelt and even dedicated "The American Presidency" to the President.[14] The admiration of some on the left for Roosevelt, and in particular his stand against fascism, was also noted by Beatrice Webb who in 1940, as war clouds hung over Europe, wrote in her diary, after hearing FDR on a radio broadcast that he was "far surpassing in decision, dignity and eloquence any other leader today".[15]

The New Deal presented less of an ideological challenge to the reformist left and those in the centre, although even here the response was not always clear cut. In December 1933 John Maynard Keynes welcomed the New Deal, with some reservations, when the *New York Times* sought his opinion on the new administration. After suggesting that the need for recovery was being overtaken by reform, he concluded "you are the only one who sees the necessity of a profound change of methods and is attempting it without intolerance, tyranny or destruction."[16] When the two men met in May the following year, neither was that impressed, and although Keynesian elements entered into the New Deal, particularly after the "Roosevelt Recession" of autumn 1937 to summer 1938, even then, as Parrish has pointed out, Roosevelt became "a Keynesian by necessity rather than conviction".[17]

The potential for a planned, mixed economy at home was picked up by other liberals, including Lloyd George, who in early 1935 presented a "Memorandum on Unemployment and Reconstruction" which recommended public works and economic planning, consciously drawing on the New Deal, but was rejected by the Government. Lloyd George was attached for a short period to the newly formed "Next Five Years Group", founded by Clifford Allen and Harold Macmillan, which argued in favour of a mixed, partially regulated economy, and centrist politics, however he had already moved on to his "Council of Action for Peace and Reconstruction," by the time the group published, *The Next Five Years: An Essay in Political Agreement* in July 1935. Among the 153 supporters listed at the front of the book were a number of literary radicals, including H.G. Wells, but the group never succeeded in attracting much Labour support, and eventually folded in 1937.[18] The Labour left were critical of such moves for ideological reasons, and the issue of unity/alliance across party lines was considered problematic following MacDonald's split from the Labour Party.[19]

Generally, the Trade Union wing of the Labour Movement was more sympathetic towards New Deal style planning than the political wing, not least because they recognised the improvements in working conditions, and related advances made by the US Trade Unions. In 1933, TUC General Secretary, Walter Citrine, told the TUC Congress that whilst "social ownership and control furnishes the only adequate and lasting solution," they were impressed "by the vigorous efforts now being made by President Roosevelt toward the stimulation and regulation of industry by means of the Industrial Recovery Act and allied legislation."[20] Citrine's comment is interesting, however, as it revealed the trade unionists distinction between short and long-term policies, and prevented a wholehearted embrace of the New Deal.[21]

It was not until the later 1930s that American liberalism became more consistently influential in the Labour Movement, when revisionists like Evan

Durbin picked up American ideas about changing patterns of ownership and control. A.A. Berle and Gardiner Mean's *The Modern Corporation and Private Property* (1932), was important in this regard, arguing in favour of a separation between ownership and management. Durbin and others argued that this presented an opportunity for reform without necessarily needing blanket public ownership. They also suggested that capitalism was here to stay and if anything had become more advanced in the US, arguments which played a role in the dispute between left and right in the party in the 1950s.[22]

At times the left's debate over the New Deal, and its applicability to Britain, was conducted in the realms of theory, with little regard for the differences between the US and Britain and the consequent difficulty in translating US political programmes across the Atlantic. There were, however, a large number of British Radicals who visited the US in the 1930s and witnessed New Deal America firsthand. They followed a well-worn lecture trail which had first been laid in the late nineteenth century. US universities, voluntary organisations, journals and newspapers seemed to have an insatiable appetite for the opinions of British intellectuals in these years; something which the latter were more than willing to supply, as the circuit paid well (especially for any Labour politician out of work after 1931). America also offered the prospect of exciting new experiences. Indeed the last factor was often vital, as the US had a fascinating allure even for those who saw themselves as anti-American.

On arriving in the US, British radicals were frequently surprised, and sometimes overwhelmed by what they saw. The were received warmly, but also discovered that a great deal was expected of them, sometimes in terms which might grate with their political views. For example, Jennie Lee called the lecture trail a "trade route between Great Britain and the States,"[23] while Vera Brittain recorded in her diary how her agent told her she owed 40 lectures in 25 cities to the organisations who had "purchased" her.[24] The sheer size and diversity of the nation (in the case of those who made it beyond the boundaries of Washington DC), was also significant, as was the discovery of democratic instincts within American political culture (visiting radicals often socialised in liberal circles, mixing with New Deal officials and supporters).

The intensity of the US visits also left many confused. Rebecca West visited the US from the 1920s, and initially thought she might settle there. She lectured, travelled and socialised on both the east and west coasts. A popular speaker, she pleased audiences, and also attracted a number of admirers on her trips—including Charlie Chaplin.[25] Despite the welcome income tours brought in, and support from US publishers, and various friends, she changed her mind about settling permanently, and indeed became more critical of what she considered violence, corruption, and the shallow moral standards of US.[26]

In a similar vein, Jennie Lee found the US equally hard to comprehend. One of a number of MPs, or former MPs who visited US during 1930s, Lee first lectured in US in early 1931, whilst still the ILP MP for North Lanark, and then made several visits after she lost her seat in the election of that year. Lee, like many other young socialists, had read Emerson and Upton Sinclair at university, which had done little to temper her strong suspicions of the US. Her visits were thus concerned as much with financial rewards on offer, as any intellectual or cultural interests, and she was marketed as the young, dynamic "girl wonder" of British politics. Once there, however, Lee found America a somewhat confusing experience. Unlike many other visitors, she travelled widely, speaking to a mixed audience of white and African-American sharecroppers in the South, closely watched by the local sheriff and his "gunmen"; and being arrested when supporting striking miners in Kentucky in 1932. She also found time to discuss America with Walter Lippman and Sinclair Lewis whilst in New York.[27] In her 1939 memoir, "Tomorrow is a New Day," the indefatigable and opinionated socialist when it came to so many other matters came over as more flexible and indecisive, commenting "I could never quite make up my mind whether this was the finest or the damndest country in the world." She went on "I totally failed to form any coherent picture of America. I don't think it can be done. And the more Americans explained America to me, the more blurred the pattern became."[28]

Of course, not everyone felt as confused by their experiences in US, and some spent longer in the country and engaged more fully with its complexity, drawing what they felt were more coherent visions of the country. However, even then different individuals had a tendency to remake their image of America to fit their own political template, and also filter it through their sense of Englishness. H.G. Wells and J.B. Priestley are interesting on this regard, as they both spent considerable time in the US, and wrote a significant amount about the country. Wells was, of course, always an idiosyncratic socialist, who based his political vision not on an anti-modern defence of the past, but an enthusiastic embrace of an organised, internationalist future, which would fulfil the best traditions of English radicalism. Significantly, unlike many on the left, Wells also had a positive view of America, believing that it represented a forward step on the path to this future.

Wells' views on America were shaped by several visits, wide reading, and correspondence, occasional affairs and debate with a large number of eminent Americans. Among the latter were Margaret Sanger, Upton Sinclair, Henry James, Booker T. Washington, Theodore Roosevelt, and even Franklin Roosevelt himself.[29] Wells first visit to the US in 1906, described in his *The Future in America*, convinced him of the primacy of America in the modern world, but left him unsure as to whether this would have beneficial or harmful consequences.[30] While his doubts were never entirely extinguished, he became more positive about America's potential, and by the end of the war, his support for Wilson, and the League of Nations, led him to believe that it would be a vital and important player in deciding the shape of the post war world—a view he, characteristically, was quite happy to make public in the most colourful terms. Thus in December 1919, he sent a message to a meeting celebrating US-British wartime co-operation. The message, subsequently reprinted in the *New York Times*, announced:

> Greetings to the great English-speaking Republic which has solved itself, and must now help the world solve its problems of diversity and unity and republican freedom with order and stability. European crowns flicker and fall like candles burned to the socket. Let there be no more of these false flares, but let star after star be added to your pioneer star until the whole human sky glows with liberty, equality and brotherhood.[31]

Wells' view of America was based on the belief, stated in his *Outline of History* (1920), that Britain, France and the US shared a common heritage of radicalism.[32] In the post war world it was the powerful US which carried the torch for this tradition; one which by the 1930s, Roosevelt seemed to personify. Although Wells initially believed that the New Deal might be defeated by the conservative opposition, his 1934 visit to the US and meeting with the President encouraged a more optimistic perspective, describing FDR as a potential convert for his "Open Conspiracy," lavishing fulsome praise on both the President and Eleanor Roosevelt. "My impression of (them) is that they are unlimited people, entirely modern in the openness of their minds and the logic of their actions...Here in the White House the unblinkered mind was in possession."[33] Even when the New Deal had run out of steam, and a more downbeat Wells made his final visit to the US in winter 1940, he still believed that radical tradition, epitomised by Roosevelt, would resurface in America and offer hope to another post war world.[34]

J.B. Priestley was also a prolific writer and a talent for forays into social criticism. Although 30 years Wells' junior, the two men became good friends, and Priestley was similarly interested in radical ideas. Indeed Priestley gave the oration at Wells' funeral in August 1946, describing him as a man "who saw further than other men did."[35] Priestley also had a strong interest in the US, inspired in part by his need to promote his books and theatre productions, as well as his appearances on the gruelling, but financially rewarding lecture circuit. As a consequence, Priestley travelled frequently to the US from the 1930s onwards. Unlike Wells, Priestley has usually been seen as representing classic English suspicion of America—something which the American press suggested before he even set foot in the country. However, as Priestley frequently pointed out, he was not anti-American, and indeed his take on America was more interesting than the latter label suggested.[36]

Priestley travelled widely on his visits, experiencing some of the social, cultural and economic diversity that America offered. His first volume of autobiography *Midnight on the Desert* described his experiences on the East and West coasts, as well as in the south West, specifically, Arizona (the latter providing the desert of the title). Generally, the book is positive and even, at times, optimistic about America. Priestley was particularly effusive about the Arizona landscape which he came to place beside the love of his native Yorkshire countryside, as well as the democratic nature of the communities he discovered there, who were "reasonably sure of themselves, easy in their minds and not galled by feelings of inferiority."[37]

He was also impressed with parts of the more developed side of America, including the large engineering feats like the Boulder Dam, which he believed represented the hint of new, collective future.[38] The Boulder Dam was completed under the auspices of the New Deal and Priestley, like many of his English counterparts, greatly admired Roosevelt, although he believed the New Deal itself, whilst an improvement, could have offered a more radical change of direction.[39] Priestley was more sceptical of other aspects of "modern" America, including the increasingly pervasive mass culture, which he believed was undermining local, vibrant, democratic traditions. Hollywood was held up as a particular example of this process. Priestley was interested in film, and occasionally wrote for Hollywood (although he kept his name off the credits), but he disliked the commercialism of the industry and the way it aimed at the lowest common denominator.[40]

Like other British radicals, he was also concerned at the way Hollywood's influence was felt outside America's borders, and this formed part of a critique, which Priestley would develop after the Second World War, of the way America's social, economic and cultural values were being exported with deleterious effects around the world. As early as 1933, his classic social critique, *English Journey* contained a number of references to the Americanisation of aspects of British life, which Priestley believed were part of the "third" England he discovered on his journey. He suggested that much like the Woolworth store which typified this England, there was greater equality and choice, but the there was also standardisation and a loss of what he described as "zest, gusto, flavour, bite, drive, originality."[41]

By the later 1940s, the acceleration of America's cultural impact on Britain, combined with Priestley's disillusionment with what he saw as the loss of the wartime collective spirit, led him to be more sceptical America and the mass society in general. The Cold War also played a role in this, and the domestic manifestations of anticommunism in the shape of the demagogic Joe McCarthy, caused a permanent rift with fellow writer Rebecca West.[42] Priestley wrote to his old college friend, the poet Edward Davison, that "it is a mistake to assume that anti-communism and Americanism are the same thing," that Western Europe shared to some extent with America a belief in "justice, freedom, truth," but that this did not mean, Priestley suggested, that "we want to turn ourselves into imitation Americans, or imagine that the American way of life is the best way of life."[43] Five years later Priestley co-authored with his third wife, Jacquetta Hawkes, the travel/social comment book, *Journey Down a Rainbow*, which coined the phrase "admass" to describe the mass society which operated in much of America and which was the "great invader" in Britain.[44]

Both Priestley and Wells therefore respected Roosevelt, and appreciated elements of contemporary America. They also both accepted that it was at the cutting edge of the modern world, although Priestley, increasingly sceptical even in the 1930s than the more idealistic Wells, was already turning against Americanised modernity by the outbreak of war. Significantly, another influential cultural figure, W. H. Auden, drew similar conclusions about America's significance, but rather than turning against this and celebrating a rooted, local culture, he moved to the US to live and work at the heart of the modern world.

Auden's departure in January 1939, with war fast approaching, caused consternation within literary and wider circles. Auden, however, was tired of English literary life, and the fame and notoriety which he and the other leftist poets had attracted. The rise of fascism and the approaching conflict were undermining his belief in the political role of poetry, as well as his wider interest in communism and he also sought a freer personal environment.[45] Auden welcomed the "atomisation" of the modern order and the "disintegration of tradition" which was most apparent in America, believing that once people accepted this they could "achieve a real unity through a common recognition of their diversity."[46] Auden wrote to his friends, the Dodds, that "owing to its newness one can see clearly in America what is true for Europe too, but is obscured there by the still vigorous past i.e. one can see what pattern of life the Machine permits." The "machine" was an important image for Auden, as it symbolised this more modern, traditionless, mobile world, where the individual stood alone. Of course this presented dangers as well. As he concluded the letter, "As an artist, I believe America to be the best place to live, because here it is impossible to deceive oneself. All you say about its destructive power on writers is perfectly true. It is a terrifying place … but the best of which one is capable is more likely to be drawn out of one here than anywhere else".[47]

Auden elaborated on these points in his major poem of 1940 "The Double Man" (published as New Year's Letter in Britain), which contained the lines

> More even than in Europe, here
> The choice of patterns is made clear
> Which the machine imposes, what
> is possible and what is not
> To what conditions we must bow
> in building the Just City Now.[48]

The poem marked Auden's engagement with metaphysics and his gradual embrace of Christianity as an alternative to Marxism.[49] Auden wrote, taught and found love in America and, whilst not blind to the country's faults, identified with it.[50] So much so that in April 1945 he flew to Europe as a US Army major, as part of the "United States Strategic Bombing Survey", which was examining the effect of bombing on the civilian population of Germany. Stopping on route in England, Auden amazed some his friends with his pro-American views, with some suggesting that he even spoke with a slight American accent (something Auden denied.)[51] Auden became an American citizen in May 1946, but his identification with America lessened as time went on.[52]

British radicals in these years, therefore, found America a somewhat confusing proposition. It was easy to dismiss on ideological grounds, but the significance of America in the modern world, and the advent of the New Deal caused some to rethink. Those who made the trek across the Atlantic also discovered a more complicated country than they had expected, and one which they often enjoyed. Those who spent longer there, like Wells and Priestley arrived at a deeper understanding of America, but had a tendency to rewrite America to fit their own vision of Radical Englishness. Auden shared their emphasis upon the importance of America, but his embrace of the atomising elements of American modernity marked a stark contrast to Priestley's localism or Wells' "Open Conspiracy." And in the years that followed, Auden's position was arguably the most appropriate. However, even here there was an element of re-imagining. The "machine" might apply to Los Angeles or New York, but in other parts of the United States, tradition played an important role. The legacy of slavery, the civil war and the disfiguring continuities of racism were one example of this, although Auden, like virtually all the other British radicals, was mostly silent on the blatantly obvious discrimination against African-Americans and other minorities. More generally, tradition, community and community operated underneath a language of individual freedom.

That having been said it is probably not surprising that British radicals should have found America hard to understand, since America as a place or an idea is a complex collection of places and ideas. In this sense Auden probably captured the feelings of many visitors and residents alike, when in his libretto to Benjamin Britten's Opera, *Paul Bunyan* which he began writing in 1939, the central character concludes that "Everyday America's destroyed and re-created, America is what you do, America is I and you, America is what you choose to make it."[53]

---

[1] My thanks to participants who commented on an earlier draft of this paper at Warwick and the Institute of Contemporary British History. Thanks also to Tim Lockley for his comments on a later draft.

[2] Henry Pelling, *America and the British Left: From Bright to Bevan* (London, 1956), 1-2.

[3] *Ibid.*, 49-65.

[4] Roger Fagge, *Power, Culture and Conflict in the Coalfields: West Virginia and south Wales, 1900-22* (Manchester, 1996), 249. Aneurin Bevan wrote about the influence of Debs, DeLeon and Jack London on himself and others in south Wales. See Aneurin Bevan, *In Place of Fear* (London, 1952), p17-18.

[5] On the British left and its changing relationship with America see, among others, Pelling, *America and the British Left*; Stephen Brooke, "Atlantic Crossing? American Views of Capitalism and British Socialist Thought", *Twentieth Century British History*, 2.2 (1991), 107-36; Steven Fielding, ""But westward, look, the land is bright!" Labour's revisionists and the imagining of America, c. 1945-64", in J. Hollowell (ed.), *Twentieth-Century Anglo-American Relations* (London, 2001); Hugh Wilford, "The South and the British Left", in Joseph Ward (ed.), *Britain and the American South: From Colonialism to Rock and Roll* (Jackson, Miss., 2003),163-86; Peter Conrad, *Imagining America* (London, 1980) provides some useful insights on cultural figures.

[6] See Pelling, *America and the British Left*, 130-1 on US economic strength after World War I. Ross McKibbin discusses the issue of Americanisation, focusing on language in *Classes and Cultures: England, 1918-5*1 (Oxford, 1998), 511-13

[7] George Orwell, *Coming Up For Air* (London, 1939), 25-6.

[8] Brooke, "Atlantic Crossing", 108.

[9] Ben Pimlott, *Labour and the Left in the 1930s* (Cambridge, 1977), 9-15.; Pelling, *America and the British Left*, 131

[10] For a good analysis of the New Deal, see Anthony J. Badger, *The New Deal: The Depression Years, 1933-40* (London, 1989). On Labour and the CIO, see Robert Zieger, *The CIO, 1935-55* (Chapel Hill, 1995).

[11] Pimlott, *Labour and the Left*, 40-2, 63.

[12] Quoted in Pelling, *America and the British Left*, 144. See 143-6 on Laski.

[13] Pimlott, *Labour and the Left*, 40.

[14] Pelling, *America and the British Left*, 144.

[15] Entry for 11 June 1940, *The Diary of Beatrice Webb, Volume 4, 1923-43* [Ed. Norman and Jeanne MacKenzie] (London, 1985) , 454.

[16] New York Times, 31 December 1933, reprinted in G.H. Bennett, *Roosevelt's Peacetime Administrations, 1933-41: A Documentary History of the New Deal Years* (Manchester, 2004), 153-4.
[17] Michael E. Parrish, *Anxious Decades: America in Prosperity and Depression, 1920*-41 (New York, 1994 edn.), 376-79.
[18] Pimlott, *Labour and the Left*, 145-7.
[19] See *ibid.*, Chapter 15 more generally on this subject.
[20] Quoted in Pelling, *America and the British Left*, 136.
[21] Barbara C. Malament, "British Labour and Roosevelt's New Deal: The Response of the Left and the Unions", *Journal of British Studies*, 17.2 (1978), 165.
[22] See Brooke, "Atlantic Crossing?", and Fielding, ""But westward, look, the land is bright!"
[23]Jennie Lee, *Tomorrow is a New Day* (London, 1939), 154.
[24] She preceded this with the suggestion that the tour 'sounded appalling" and she "wondered not whether I should break down but when.", Vera Brittain, *Dairy of the Thirties, 1932-39: Chronicle of a Friendship* [ed. Alan Bishop] (London, 1986), 171-2.
[25] On one occasion Chaplin and West were apprehended after breaking into the Central Park Boathouse, and were only allowed to leave after Chaplin had performed some of his routines. Carl Rollyson, *Rebecca West: A Saga of the Century* (London, 1995), 74-5.
[26] On West and America , see *Ibid.*, 72-80 , 86-7, 103-4, 141-2, 226, 247.
[27] Lee, *Tomorrow is a New Day*, 156-163. Patricia Hollis, *Jennie Lee: A Life* (Oxford, 1997), 74-5.
[28] Lee, *Tomorrow is a New Day.*, 162
[29] See H.G. Wells, *An Experiment in Autobiography, Volume. II* (London, 1994 edn., orig. London, 1934), 755-8, 784-98; Michael Foot, *H.G. The History of Mr. Wells* (London, 1995), 67-70, 190-1.
[30] H.G. Wells, *The Future in America: A Search After Realities* (London, 1906).
[31] H. G. Wells, *The Correspondence of H.G. Wells, Volume 3* [ed. David C. Smith] (London, 1998), 21-2.
[32] Foot, *H.G.,* 184.
[33] Wells, *An Experiment in Autobiography, Volume. II*, 784-98. Quote, 796.
[34] Foot, *H.G.*, 285-90. Wells' relationship with America was given an added dimension when, on 30 October 1938, Orson Welles' broadcast an American version of "The War of the Worlds", causing panic across parts of the nation, who believed that the broadcast was describing actual events. Wells was originally angered by the broadcast, but subsequently got on with Welles. Jeanne and Norman Mackenzie, *The Time Traveller: The Life of H.G. Wells* (Worcester, 1973), 410-11; Foot, *H.G.*, 289.
[35] Printed funeral oration, (File 2, J.B. Priestley uncatalogued papers, J.B. Priestley Archive, Harry Ransom Humanities Research Center, University of Texas at Austin).
[36] Priestley gave an interview as his boat docked to unload cargo in Boston, when he first visited the US in 1931. The subsequent report suggested that Priestley had been highly critical of the US, something the author vehemently denied. However, Priestley believed the incident helped shape America's reaction to him for years to come. J.B. Priestley, *Midnight on the Desert* (London, 1935), 18-21. Priestley would deal with the Anti-

American label at various times in his career. See, for example, Priestley, "Who is Anti-American", *New Statesman*, 5 Nov 1955, reprinted in Priestley, *Thoughts in the Wilderness* (London, 1957), 148-54.

[37] J.B. Priestley, *Midnight on the Desert* (London, 1937), 101-2. On landscape, see 90-4.

[38] *Ibid.*, 110-11

[39] Roger Fagge, "J.B. Priestley, the "Modern" and America," Anglo-American special issue of *Culture and Society* (forthcoming, 2007).

[40] Roger Fagge, "From the Postscripts to Admass", *Media History*, 12.2 (August 2006), 104.

[41] For Priestley's discussion of the different Englands, see J.B. Priestley, *English Journey* (London, n.d., orig., London, 1933), 393-406. For his description of the Americanised third England see, 401-6. Quote, 405. Priestley's critique of the impact of Hollywood and other aspects of mass culture was also apparent in his fictional writing. See for example J.B. Priestley, *Angel Pavement* (London, 1930).

[42] See J.B. Priestley, *The Secret Dream: An Essay on Britain, America and Russia* (London, 1946) on his misgivings about the cold war. On the disagreement with West, see Rebecca West to Priestley, 22 June 1955, (Box 2, Folder-Recipient W-Z, J.B. Priestley Archive)

[43] Priestley to Davison, 13 August 1950, (Folder 3, J.B. Priestley Correspondence of Edward Davison, 1922-70, Beinecke Library, Yale University).

[44] J.B. Priestley (with Jacquetta Hawkes), *Journey Down a* Rainbow (London, n.d., orig. London, 1955) xi

[45] Richard Davenport-Hines, *Auden* (London, 1995), 176-81. Kathleen Bell, "A Change of Heart: Six Letters from Auden to Professor and Mrs. E. R. Dodds, Written at the Beginning of World War II", in Katherine Bucknall, and Nicholas Jenkins (eds.), *Auden Studies I: W.H. Auden: "The Map of all My Youth". Early Works, Friends and Influences* (Oxford, 1995), 96-115. Interestingly, Priestley was not convinced by Auden and his counterparts, telling Davison that they had "brains" but were "a little cold and conceited." Priestley to Davison, Folder 2, Davison Correspondence.

[46] Quoted in Bell, "A Change of Heart", 99

[47] Letter 4, reproduced in ibid., 109-11.

[48] W.H. Auden, *Collected Poems* [ed. Edward Mendelson] (London, 1976), 238.

[49] For a perceptive interpretation of the poem, see John Fuller, *A Reader's Guide to W.H. Auden* (London, 1970), 131-47.

[50] Humphrey Carpenter, *W.H. Auden: A Biography* (London, 1983 edn., orig London, 1981), 282.

[51] *Ibid.,* 333-4.

[52] For Auden after the war, see *ibid.*, 336-455; Davenport-Hines, *Auden,* 228-344

[53] Libretto to Britten, *Paul Bunyan* reprinted in 1988 Virgin Records recording, VCD 7 90710.

CHAPTER ELEVEN

"A TUNNEL BETWEEN TWO WORLDS": IMAGINING AMERICA IN BRITISH WRITING OF THE 1940S

ALLEN MCLAURIN

The main chronological focus of this paper is the 1940s, a period in which we might expect issues around the question of British national identity to be especially urgent and therefore explicitly addressed in the writing of the time. It would also be surprising if an imagined America were not in some instances identified, positively or negatively, as a "significant other"; it was, after all, during this decade that the idea of the "American century," first propounded earlier in the century, seemed to become an unavoidable fact. The alliance between the two countries during the Second World War, and the great popular admiration in Britain for President Roosevelt, further cemented the bond. In the post war period, however, this positive relation was subject to considerable pressure. Britain's subordinate position, masked during the war by Churchill's dream of a continuing British Empire, became clearer; economic circumstances created dependency on large British loans from the U.S. through the Marshall Plan. From the American perspective, the election of the Labour Government in 1945 was treated in some quarters with great suspicion.

Some of the most telling analyses of the way in which America figured in the British imagination during the 1940s and 1950s have concentrated on the idea of Americanisation, a term which almost invariably carries with it negative connotations. The work of Richard Hoggart, F.R. Leavis, George Orwell, and others, is frequently referred to in such discussions as representative of the influential view that American popular culture might undermine British national identity, or at least pose a threat to an indigenous working-class culture. It is true that a number of prestigious texts from the mid-twentieth century could be cited as evidence of a "negative consensus" around the question of Americanisation. The convergence between such ideologically disparate writers as Orwell, Leavis, and Hoggart on this issue is an indication that we are dealing here with a rich discursive formation that may well constitute a kind of mainstream. There has therefore been an understandable emphasis amongst more recent cultural analysts on this aspect of British understanding, or misunderstanding, of America in the 1940s and 1950s, alongside an interest in the emergence of resistance to these dominant negative attitudes. The discussion which follows begins with a consideration of Dick Hebdige's essay *Towards a Cartography of Taste 1935-1962* which has been very influential in establishing this approach.[1] However, it is possible to speculate that his focus on "Americanisation" both illuminates and obscures the question of how America was imagined. For example, a text such as Evelyn Waugh's *The Loved One,* although it contributes to this "negative consensus" about America, does so from a position only obliquely related to the idea of Americanisation. Nevertheless, this influential novel clearly needs to be included in any consideration of the way in which America figured in the British writing of the 1940s. It is also worth pointing out that there are other, lesser-known, texts standing "out of sight of the discourse of Americanisation." but which can be examined for the more positive ways in which they represented America to their British readers.[2]

The texts to be examined in more detail are not considered in chronological order. Waugh's novel (perhaps novella would be a better description of *The Loved One*) is considered before the others, although written later. This is because it links directly to Hebdige's argument and a consideration of this text points up the difference between the angle taken in this paper and those approaches which remain within the discourse of Americanisation. Another reason for taking it first is that in contrast with the other two texts it represents a backward-looking view. It was written in the late forties, at a time when Britain was financially most deeply indebted to America, through the implementation of the Marshall Plan. Yet it embodies what might be seen as a regressive fantasy in which a new kind of post-Imperial British adventurer goes to America and brings back rich experiences which can be turned into artistic gold. The authors of the other works to be examined, Louis MacNeice and Henry Treece, in their different ways, point forward. MacNeice hoped for a new political order to be established after the war, more soundly based than traditional British empiricism, yet maintaining the sovereignty of the individual.[3] Treece in his essay writes about jazz and the way in which he relates it to classical music prefigures the erosion of the barriers between "high" and "low" culture which was to become a central feature of cultural debates later in the century.

In his wide-ranging essay *Towards a Cartography of Taste 1935-1962* Dick Hebdige established the terms in which Anglo-American cultural perceptions were to be discussed in many subsequent analyses. In his argument, Leavis, Orwell and Hoggart, with different levels of complexity, all resisted a perceived "levelling down" which came to be associated with Americanisation. In a section amusingly entitled *Milk bar horrors and the threat of youth* he quotes aptly from two texts which straddle the period under discussion. *The Uses of Literacy*, which Hoggart began to write in the early 1950s and published in 1957, illustrates the theme perfectly. It contains a generic description of cafes where slouching "juke box boys" congregate:

> Many of the customers—their clothes, their hair-styles, their facial expressions all indicate—are living to a large extent in a myth-world compounded of a few simple elements which they take to be those of American life.[4]

A decade earlier George Orwell had described one of these infamous milk bars in his novel *Coming Up for Air:*

> There's a kind of atmosphere about these places that gets me down. Everything slick and shiny and stream-lined: mirrors, enamel and chromium plate whichever direction you look in. Everything spent on the decorations and nothing on the food. No real food at all. Just lists of stuff with American names, sort of phantom stuff that you can't taste and can hardly believe in the existence of.[5]

The reference to streamlining in this passage allows Hebdige to make a neat link to the final part of his article in which he argues that the words "America," "jazz" and "streamlining" set in motion a number of ideologically charged connotational codes. The eventual displacement of one kind of negative discourse by a more positive oppositional one is in the last analysis brought about by the appearance on the scene of a new range of commodities "emanating from a zone which is 'in the last instance' more decisive and determining than language—the sphere precisely of production."[6]

Hebdige's analysis has proven to be very persuasive, and a number of writers have built upon his work. These cultural analysts generally follow his lead in constructing a narrative of fearful anti-Americanisation (exemplified in the writings of Hoggart and Orwell) which is in turn challenged by an oppositional appropriation and reworking of American popular culture. In more recent times the argument has moved on as the theme of Americanisation is displaced by a new focus on globalisation. Cresswell and Haskin, for example, argue that the idea of "American" popular music needs to be deconstructed since popular music as part of a globalised industry cannot be as easily be mapped on to nations.[7]

Despite this more recent change of emphasis, the idea of "Americanisation" has provided a powerful and illuminating focus for a narrative about British attitudes towards America in the middle of the twentieth century. But it is possible that the wider and more general question of how America was imagined will lead to different readings, and consideration of other texts. A case in point is the use that Hebdige makes of the writings of Evelyn Waugh's writings. Hebdige's starting point in his essay is not Orwell or Hoggart, but Waugh, whose works "codify the rituals of exclusion through which the power of an already superseded ancien regime had been perpetuated."[8] A perfectly defensible statement, exemplified with reference to Waugh's characterisation of Trimmer, a working class upstart on the make who appears in the novel *Officers and Gentleman*. However, there is no reference at all to Waugh's best-known writing about America; Hebdige does not allude to Waugh's fictional representation of America in *The Loved One*. The novel is not only explicitly about the U.S., it also has the relationship between high and popular culture as a major theme, and so to some extent overlaps with the discourse of Americanisation. For this reason, and because it is probably the most familiar literary work of the 1940s that relates to the topic in hand, it is worth looking in more detail at the context in which it first saw the light of day, and the way in which it constructs an imaginary America.

The Loved One was first published as a special issue of the magazine *Horizon*, which was edited by Cyril Connolly, and there is a jocularly knowing reference to this literary journal in the novel itself. *Horizon*, which flourished in the 1940s, championed Modernism, and showed a particular interest in avant-garde American painting and poetry. In relation to the high versus popular culture argument, which is closely associated with attitudes towards and representations of America during this period, *Horizon* seemed to be ideologically ambiguously placed. Brookeman in his study of the "moment" of *Horizon* points out that:

> …Connolly hedged his bets in the late 1940s by also foregrounding American painting and poetry in a number of special issues that represented another important wave of the future. There is also an excellent piece of early McLuhan on American advertising that takes a more positive view of mass culture than was the case with Greenberg's early scathing attack on the kitsch products of mass culture…It was out of this shared Anglo-American thinking that the negative view of mass culture was constituted…These opinions were circulated and refined in the many discussions that intellectuals initiated about the role of high culture and mass culture throughout the 1940s and 1950s [9]

It was in this interesting context that Waugh's *The Loved One* was first read. The novel follows the adventures of a sardonic English poet, Dennis Barlow who works in The Happier Hunting Ground, a funeral parlour for pets in California. Its counterpart for human beings, the Whispering Glades Memorial Park, is the workplace of Aimee Thanatogenos, and the story follows Dennis's attempts to seduce Aimee by passing off selected passages from the Oxford Book of English Verse as his own work. Throughout the novel, the cosmetic approach to death is intended to epitomise the lack of any true religious or spiritual feeling in American society. Cyril Connolly, wrote a brief Introduction in which he quoted Waugh's enumeration of his intentions in writing the novel:

> The ideas I had in mind were: 1. Quite predominantly, over-excitement with the scene [the cemeteries of Southern California]. 2. The Anglo-American impasse—"never the twain shall meet." 3. There is no such thing as an "American." They are all exiles uprooted, transplanted and doomed to sterility. The ancestral gods they have abjured get them in the end. 4. The European raiders who come for the spoils and if they are lucky make for home with them. 5. *Memento mori.*[10]

Waugh predicted that the novel would cause "ructions" but flattered the *Horizon* readership by observing that he had chosen to publish in that journal because its readers were "tough stuff." Although Connolly was extremely complimentary—"one of the most perfect short novels of the last ten years"—it is possible to detect a certain unease in his introductory remarks. He attempts to appease his American readership by suggesting that Waugh's "Anglo-American impasse" should be replaced by "Anglo-Californian impasse." But then he qualifies his initial demurral by saying that Southern California is ten years ahead of the rest of the country and so the novel is about America after all, exposing a "materialist society at its weakest spot."[11] This switch from the national to the local (and back again in this case) is a significant move in any representation of national identity.

The novel is in part a response to a transitional moment when in Britain a sense of the inevitable loss of Empire was becoming clear. It has a "trick" beginning, with the opening scene presented as though the setting were some god-forsaken outpost of empire: "the dry sounds of summer, the frog-voices, the grating cicadas, and the ever present pulse of music from the neighbouring native huts." The scene continues with two English men, "the counterparts of numberless fellow-countrymen exiled in the barbarous regions of the world," drinking their sundowners on a veranda next to a "dry water-hole."[12] Only after the first page or so is the scene revealed as being in America. Everything in Hollywood, a place dedicated to the creation of images, is a simulacrum, and Waugh aptly begins the novel by deceiving his readers with an imitative form which mimics cinematic illusion. But the effect of this opening deception goes beyond trickery; it portrays the real sense of disorientation that must have been felt by many British people at the demise of the British Empire, and a dawning realisation of American world supremacy. However, *The Loved One* is not an expression of twilight nostalgia, it seems to be more like a consoling fantasy, with Waugh regarding himself, like his protagonist Dennis Barlow, as one of those "raiders who come for the spoils,"[13] taking back from America "the artist's load, a great, shapeless chunk of experience; bearing it home to his ancient and comfortless shore."[14] The "raider" idea prompts a reading of the satire as an act of revenge against the US as the "usurping" power. This revenge also takes another form, the feminisation of America implicit throughout the novel, and especially in the characterisation of Aimee. She is a creation designed to show that there is "no such thing as an American." In embodying that perception she represents the "nonentity" which is America itself in Waugh's eyes. Poor hapless helpless Aimee might incidentally be seen as Waugh's counterweight to the Statue of Liberty, the female colossus that figured in popular representations of America in Britain during the War years.

A central theme in *The Loved One*, of high culture and its "debasement," is particularly pertinent to this essay. British culture is largely defined in the novel through its literary tradition. Its superiority to American culture is established in a negative way through the ignorance of British literary culture displayed by Americans and by the grotesque travesties of it for which they are held responsible. A memorable example of this is the recreation in Whispering Glades of Yeats's Lake Isle of Inisfree, complete with the humming of electronic bees. The satire (and this is characteristic of the novel as a whole) goes beyond its immediate target—Yeats's poem itself, as well as the farcical American "materialist" approach to it is also slyly mocked. This becomes more obvious by its juxtaposition with the fake Scottishness of the Lover's Seat. With its accompanying inscription from Robert Burns, this is a concrete embodiment of the way in which, in Waugh's view, cultures do not travel. The "Balts and Jews and Slavs" who mouth the words uncomprehendingly illustrate one of the objectives in writing the novel which he enumerated to Connolly, that is to say, to show that the melting pot of America results not in what might now be regarded as a positive hybridity, but in "sterility." Again, the satire overflows its boundaries, to mock not only American credulity, but also any attempt at the periphery (the "Celtic fringe") to create an alternative to Anglo-centric "Britishness." This reading is reinforced by Dennis's "natural abhorrence of dialect" which "had prevented him from borrowing any of the texts of his courtship from Robert Burns." In order to prevent, as a response to his "there is no such thing as an "American," the *tu quoque* "there is no such thing as an Englishman," Waugh needs to mock out of existence any residual or peripheral "un-English" elements within the confines of Dennis's "ancient and comfortless shore."

It is truc that wc can read this aspect of *The Loved One* with an awareness of multiple ambiguities, and none of the characters, including the British, escape from Waugh's satirical treatment. However, the superiority of British culture, somewhat ambiguously evoked in the story line, characterisation, and literary allusions, is less ambiguous in the ironic, florid suavity of the narrative voice, which, juxtaposed as it is with the absurd rhetoric of Whispering Glades, forms the ultimate point of contrast between English and American culture. The following passage from the novel presages Aimee Thanatogenos's suicide:

> Attic voices prompted Aimee to a higher destiny; voices which far away and in another age had sung of the Minotaur, stamping far underground at the end of the passage; which spoke to her more sweetly of the still Boetian water front, the armed men all silent in the windless morning, the fleet motionless at anchor and Agamemnon turning away his eyes; spoke of Alcestis and proud Antigone.[15]

One way of reading this high-flown rhetoric would be to see it as mock-heroic satire in which Aimee becomes a modern-day equivalent of Pope's Belinda in Pope's *The Rape of the Lock*. But is the tone quite as unambiguously comical as that would imply? On another level, this passage precisely illustrates Waugh's idea, reported in Connolly's Introduction, that "there is no such thing as an 'American.' They are all exiles uprooted, transplanted and doomed to sterility. The ancestral gods they have abjured get them in the end." Where, however, do the ancestral voices come from that prompt Aimee? Clearly not from her education, which is presented as consisting of a subject, Beauticraft, made to seem all the more farcical by its accompaniments, Psychology and Chinese. The reader must assume, then, that the Greek allusions are racial memories rather than Aimee's conscious awareness of an intellectual tradition. Although the notion of "ancestral gods" elides the distinction between nature and nurture, we are forced to conclude, that like so many explorations of national identity, Waugh's representation of this issue comes down to a question of genetic inheritance, and is dangerously close to a notion of racial "purity." Further, the reference to "sterility" places Waugh's statement in very bad company within the complex discourse of hybridity, as described by Hutnyk:

> The criticisms of hybridity can be collected into several categories: the heritage of hybridity's botanical roots...the sterility of the hybrid mule, and its extension to mulatto, mixed race, half-breed and other obscene racisms; the reclamation of the term reconfigured as creativity at the margins and as advent of vibrant intersections that cannot be otherwise incorporated; the hegemony of the pure that co-constitutes the hybrid; the inconsequence of hybridity in the recognition that everyone is hybrid, everyone is "different"; the commercial co-option of multiplicities; and that if everyone is hybrid, then the old problems of race, class, gender, sex, money and power still apply.[16]

We can explore the issue of national identity further through the character of Waugh's protagonist, Dennis Barlow. There is a degree of uncertainty about the stance the reader is supposed to take towards him; this was a cause for concern in some earlier reviews and has puzzled many subsequent readers. For example, in a perceptive review, Olivia Manning (writing as R.D. Smith) concluded that:

> We feel a similar contradiction between what Mr Waugh intends us to understand are his values, and what the feel and texture of his writing reveal. Satire requires less erratic values than Mr Waugh's style and characters suggest...[17]

But this valid perception by an early reviewer is for the modern critic a starting point rather than a final judgement, as it raises the further question of the ideological function of this "erratic" satire: what do these apparent inconsistencies reveal? The reader is amused by Dennis's mordant wit, but by creating such an unscrupulous focal character, Waugh manages to undercut the idea of an unequivocally admirable contemporary Englishness against which a grotesque America can be judged. In this respect Waugh is following a familiar pattern in projecting anxieties about British identity onto a representation of America. We might turn the tables, and without the racial mystification which can be detected in his portrayal of Aimee, ask what voices are prompting Waugh. His urbane narrator seems to speak, in a tone hovering between the heroic and the mock heroic, to an older, perhaps late-Victorian or Edwardian classically-trained Englishman, as a shadowy "implied reader" – one who would appreciate the slightly arch rhetoric of such a passage as that quoted above, with its dying fall, so redolent of the 1890s.

Waugh's tale was a popular success when printed in book form (sufficiently so for David Low to allude to it in his cartoon about Truman's unexpected victory over Dewey in the Presidential elections of 1948). The work was also adapted for the cinema, with a screenplay by Isherwood, who added a further layer of irony to the satire on Hollywood contained in the novel by creating a travestied version of Waugh's original. The novel became a point of reference in discussions of Anglo-American cultural differences in the decades following its first publication. As late as 1990 Christopher Hitchens in *Blood Class and Nostalgia* found Waugh's fiction relevant to his discussion of the "Greece to their Rome" analogy which is a recurring metaphor in the elaboration of the "special relationship" idea. He quotes Harold MacMillan's version of this theme:

> We are... the Greeks of the Hellenistic Age: the power has passed from us to Rome's equivalent, the United States of America, and we can at most aspire to civilise and occasionally to influence them.[18]

Waugh represents for Hitchens a satirical version of this, prefiguring later decades of the century when the British would exert a subtle and devious influence on the Republic. Hitchens' analogy is rather strained, and is in danger of being confused by the actual Greek theme in the novel portrayed through the figure of Aimee Thanatogenos. However, his use of the tale as a frequent point of reference in his book of "Anglo American Ironies" points to the durability of Waugh's imaginary America.

So far, this discussion supplements and largely reinforces the views expressed in what has come to be a critical consensus around the Hebdige thesis, that the significant mid-twentieth century British discourse about America, prior to the rise of oppositional voices in the 1950s, was negative. (Though my approach pays more attention to some of the mechanisms by which those views were given persuasive form, and also reveals ambiguities of stance and tone.) What follows is a discussion of two further pieces of writing from the 1940s which problematise this negative view. Both examples are non-fictional prose writings and both, like *The Loved One*, originally appeared in literary/artistic miscellanies, a form of publication characteristic of the period. They also both, incidentally, happen to be written by writers better known in the period for their poetry than their prose. The first is Louis MacNeice's contribution to a series of documentary writings by various authors called *The Way We Live Now* and which appeared in *Penguin New Writing* in 1941. It has been chosen because it explicitly addresses the issue of national identity. It does so through a "traveller's return" narrative which shows how a complex perception of the relationship between American and British realities and unrealities was imaginatively expressed at a particular historical moment. The essay also demonstrates some of the complications involved in constructions of national identity in this period, and so makes problematic the "British" in "British imagination." The second is an essay by Henry Treece entitled *Henry Goes Honky Tonk* published in 1945 in a periodical called *Modern Writing*. It raises explicitly the issue of popular and highbrow culture, but from a more sympathetic angle than that which has come to be associated with the idea of "Americanisation." It therefore puts into question the notion of a straightforward pre-1950s "negative consensus" about America and popular culture which seems to have become the orthodox assumption about this issue.

There is a degree of continuity between attitudes expressed in the 1940s and earlier traditions of British writing about America. One of the major sources for an answer to the question of how the British in the past have viewed America has come from writers who have reported on their experience of undertaking lecture tours of the States, a tradition stretching back to the 19th century. Dickens and Wilde are the most often-quoted examples. A neat parallel this: Dickens experiencing culture shock, and Wilde, it appears, delivering it. Although there is evidence of changing responses to America over time, and such fluctuations are perceptible even during a restricted time period such as the 1940s, there are also strikingly persistent ideas and attitudes. Numerous British observers, for example, from the nineteenth century onwards, have commented on the prevalence and social acceptability in the U.S. of spitting in public. Max

Berger amusingly points out that although spitting was disapproved of by nineteenth-century British travellers in America, the "accuracy with which spitters hit their targets could not but evoke admiration."[19] We can see the durability of these observations and attitudes epitomised in a letter published in the Evening Standard, written by an American in response to a Mass-observation survey of British attitudes towards Americans. In the course of the letter this correspondent lists things Englishmen say:

> 1. We Americans spit all over the place. Dickens said that years ago… Now even if visitors do not see spittoons they mention the spitting habit anyway.
> 2. American hotel rooms are too hot
> 3. Our tea bags are terrible—it is something new for them—therefore it is not done.

Louis MacNeice's essay "The Way we live now" is an example of a more metaphorical response to this cultural exchange. His account concerns his return journey across the Atlantic in 1941 after a period of lecturing in America. For MacNeice setting out on his journey, the closed deck of the ship was a "black corridor," the passage across the Atlantic a "tunnel between two worlds" and the war itself a tunnel at the other side of which might be a "more concrete kind of socialism" which "will not substitute for a ruling class of money a caste of political power-addicts." [20] Consideration of this strand of imagery suggests a more general conclusion, that in relation to the issue of the way in which America is imagined in Britain, resonant images, as opposed to arguments or opinions, might represent our best evidence. [21]

One of the larger narratives about this period, from the 30s through to the post-War welfare state consensus sees a movement from the Slump and Appeasement of the thirties, Auden's "low dishonest decade," through the crisis of the war years, and culminating in the planned, more egalitarian society which would be established after the War.[22] To some degree, MacNeice's article, like a good deal of writing in the 1930s and on into the war years, can be seen as partaking in this forward-looking narrative. Though this optimism in MacNeice's case represents a fairly sudden reversal which happened around the period 1939-40, and is tempered by a concern that new developments should not lead to the "cult of the state."[23] It is clear that MacNeice's trip to America, and his homeward journey on a ship packed with passengers of many different nations, which he recounts in this essay, crystallises his view of London, and by extension, of Britain and Britishness. Alan Munton, in an analysis which relates MacNeice's poetry to his personal relationship with an American writer at this time, remarks that "It was America that gave him the personal and political experiences out of which he imagined how England might be politically transformed."[24]

In terms of narrative form the article has a number of the characteristic features of a travelogue, but as it describes only a return journey, it is an interesting variant. [25] In its own way the essay illustrates the view that "it is particularly in relation to the trope of travel that transatlantic literature explores the constructions of national identity."[26] The outset of any journey might be seen as a kind of liminal moment. Such moments have long been recognised in anthropological writings, and have informed more recent theoretical work on travel writing. In this way, a transcultural "in between" has been identified as a generic trait of travel writing. At such a threshold, proximity and distance between the foreign and the home culture can be seen in tension. The dark tunnel image articulates this liminality and it also constitutes a response to a question posed by any account of travel, as to whether the journey represents a connection or a separation. For MacNeice the return to Britain was an escape from a sense of being trapped between two unrealities. This needs some explanation. In the American press he found what he regarded as ludicrously exaggerated descriptions of the blitz. This gave him a sense of the unreality surrounding him in America. On the other hand he knew that England had undergone major changes during his ten-month absence, a transformation about which he could only speculate. Trapped between two unrealities, he felt that he needed to go through the dark tunnel to get in touch with reality, which he thought would be achieved only by his physical presence in Britain, or more specifically, and significantly, in London.

MacNeice himself personifies some of the difficulties in talking of a straightforward *British* response to America. The essay provided him, under the pressure of the war, an opportunity to declare his sense of a mixed identity :

> I have never really thought of myself as British; if there is one country I feel at home in, it is Eire. As a place to live in or write in I prefer the USA to England and New York to London. But I am glad to be back in England and, in particular in London. Because London since the Blitz has become more comprehensible. Because this great dirty, slovenly, sprawling city is a visible and tangible symbol of freedom...[27]

In expressing a preference for New York it is significant that MacNeice appears not to experience any of the discomfort felt by many left wing writers in the post-war period in expressing a positive attitude towards America. Henry James talked of the "complex fate" of being American; MacNeice's article is evidence, if any were needed, that being British, in his case, an Ulsterman, not only is equally complex, but certainly by the mid twentieth century at least, involved the idea of both Ireland and America as "others" in the construction of British national identity.

The slippage from USA to New York and from England to London in this passage is also of significance in looking at the way in which America is imagined—it goes beyond the use of the rhetorical figure of synecdoche, it is not simply a case of the part representing the whole. There is a homology between the dyad nation-city and MacNeice's political views about the collective (the nation) and the individual (the city). This line of thought about individualism connects with the anarchism which he thought characteristic of British perceptions: "the typical Englishman contains, paradoxically, an anarchist."[28] In the context of the blitz, there is something else going on here as well, which bears on the question of how wars or other assaults on cities, affect the way in which the relationship between nation and city is articulated. Looked at from the perspective of the documentary genre, MacNeice's journey follows a familiar pattern, taking us from an America which dematerialises before our eyes in the impressionist picture of New York in the opening paragraph, through to the traveller's perception, at the journey's end, of the dirty realism of Mersey docks. It is important, however, to differentiate this kind of writing from factual reportage—a distinction which can be reinforced with reference to MacNeice's contemporaneous poem "The Newsreel" in which he declares that "fact is a façade."

MacNeice had published two other articles on his American visit in the months immediately prior to this, entitled "Traveller's Return" and "Touching America" in the February and March editions of *Horizon*. They are "opinion" pieces and offer an interesting comparison with the more metaphorical non-fictional prose of the Penguin New Writing essay. MacNeice's aim in the second of these essays, "Touching America," was to criticise what he saw as ill-informed European generalisations about the USA. His attempt to create a more balanced picture contrasts directly with Waugh's caricatured vision, and is clear evidence that the anti-American consensus amongst British intellectuals in the 1940s, described by Hebdige and others, was not universal:

> Thanks to Hollywood, we think of the USA as a whirl of millionaires, hard-boiled business men, simple-minded toughs and free-living blondes—all of them housed in chromium. Thanks to such American literature as comes our way (lots of it fails to) we think of the typical American as a double-dyed materialist (in the popular sense of the word), very smart in practical affairs and very dumb in everything else, giving and taking hard knocks but quite insensitive. Streamlined extraverts. When you arrive in that country, however, you meet innumerable Americans as neurotic as yourself, even more thin-skinned and certainly no less concerned with spiritual values. [29]

The function of "streamlined" in this passage is to construct a picture of Americans that MacNeice wishes to show is a false caricature. Far from exemplifying Hebdige's thesis about the negative attitudes of British intellectuals towards America in the 1940s, MacNeice here possibly anticipates one aspect of Hebdige's critique, at least in showing sensitivity to the ideologically charged nature of "streamlining."

Since the 1920s one type of sound heard through the dark tunnel of the Atlantic has been jazz, and its reception has been an important feature in the way in which the British have perceived or imagined America. Many of the connotations of "America" for the British overlap with those associated with jazz: the modern as opposed to the old fashioned; improvisation rather than established forms; popular rather than elitist culture. Some British writings about jazz in the 40s and the later post-war period offer interesting insights into the construction of imaginary Americas. Peter Bailey, for example, writes about the "vivid imaginary presence of the popular culture of the United States of America" in 1950s Coventry, where, as an adolescent jazz pianist he went under the name of Porridge Foot Pete. He describes the "various Americas" of jazz, and more generally, how for the young, "American film and music projected a countervailing pulse of modernity that held the promise of freedom, adventure and pleasurable disorder." This was true for him in Coventry, the post-war boom town, "the Klondike city of the Midlands." [30]

The precursor 1940s text which I would like to draw attention to here is an essay by the poet Henry Treece, a well-known literary figure in the 1940s, associated with the so-called New Apocalypse movement in poetry, who later became better known as a writer of children's historical fiction. Most of Treece's poetry did not fare particularly well even when it was first published, and the New Apocalypse movement with which he is associated, stands self-condemned in his own description of it as being "organic…with all the madness and sanity of a bowel movement."[31] However, Treece wrote an interesting essay on jazz which is of significance in relation to the themes of this paper. Unfortunate though the analogy used to describe New Apocalypse poetry may appear to be, it illustrates clearly the connection between Treece's poetry (according to his description of it) and his ideas on jazz, through the shared notions of naturalness and spontaneity. It should be pointed out however that his valorising of these qualities is balanced by a highly technical description of the sophisticated nature of the music which he is discussing. Treece's jazz article was published in *Modern Reading*, a serial publication edited by Reg Moore, which had something in common with Penguin New Writing and other literary miscellanies published in the 1940s. The essay, entitled *Henry Goes Honky Tonk,* begins with a highly technical discussion of the formal features of a certain type of jazz, and evokes what Peter Bailey calls the "pleasurable disorder" associated with this kind of music.[32] Treece then goes on to conjure up an imaginary America that draws on familiar stereotypes:

> These men and this music mean America to me; not the America I know when I …read Robert Frost, but an America, nevertheless, and a very valid one; an America of smoke-laden, below-stairs dives where a thug talks quietly out of the side of his mouth to a moll in a fur coat, and a third-rate pugilist argues in a Bowery voice with a wild-collared newspaperman …[33]

Treece recognises that there are various Americas and it is clear from his description that they are for him textual creations. In this passage jazz evokes a melange of quite familiar stereotyped images drawn, one would guess, from film representations of 1920s and 1930s America or possibly from hard-boiled detective fiction, or both. This is America seen through the spectacles of media texts of various kinds, and jazz placed in what was a familiar association with criminality. On one level, a perfect example of Americanisation, reinforcing a commonly held view that in an age of U.S.-dominated mass media, America itself produces the images by which it will be imagined. However, there is a playfulness, a conscious fictiveness, in this creation of a kind of urban pastoral, which makes the idea of cultural imperialism seem rather heavy handed.[34]

Part of the interest of Treece's essay lies in the explicit and detailed way in which jazz is placed in relation to high culture. Treece liked the evocation of people and places evident, for example, in the titles of some of the jazz pieces he admired. He is thinking of titles such as *Chicago on my Mind*, as the inspiration for his own essay title, *Yancey Goes Honky Tonk*. The unselfconscious way in which he compares this to the effect of some titles in Tudor music is typical of the easy movement between high and popular culture throughout the piece. The capacity of jazz to bring together "high" and "low" culture had been noted a decade earlier by Constant Lambert, as McKay points out in his excellent study of the cultural politics of jazz in Britain:

> The apparently democratic blurring and subsequent merging of high and low culture which would challenge the hierarchy of cultural value was recognised quite early on by Constant Lambert as part of the appeal of jazz: "It is the first dance music to bridge the gap between highbrow and lowbrow successfully." The high-lowness is often identified as a feature of American culture more generally….[35]

Treece was a classically-trained pianist who later worked for a while in a jazz band. Although somewhat ambivalent in his attitude towards this popular music as a high art form, he treats jazz with great seriousness, huge enthusiasm, and crucially, he claims for it greater emotional power than classical music and other classical art forms. This theme in Treece's essay casts an interesting sidelight on discussions in later decades, within cultural and literary studies, about attitudes towards high and popular culture. As we have seen, these stress negative attitudes amongst intellectuals towards American popular culture in Britain during the 40s and 50s.

The association between juvenile delinquency and American popular culture is one of those chains of association identified as a feature of cultural criticism in the 40s and 50s. In a number of ways Treece's essay prefigures, though from a contrary perspective, some of the issues raised about American culture in Richard Hoggart's *The Uses of Literacy.* Hoggart explicitly excludes any discussion of jazz, though he gives no reasons for his exclusion. We can only speculate that this may be because some of the ambiguities in relation to the status of jazz, which we have been tracing, made it difficult to fit into his polemic about the deleterious influence of American popular culture in Britain. Hoggart's book is full of a familiar moral anxiety at the influx of American popular culture. He evokes images of America, drawn from hardboiled fiction, which bear some resemblance to Treece's but they are for Hoggart entirely negative rather than positive, and there is a clear implication in some passages that American popular culture and delinquency are closely allied. Quoting Dr Rodzinsky, a member of the New York Philharmonic, Treece raises the spectre of a connection between jazz and delinquency. However, the crucial difference is that he does so only in order to cite evidence against this theory, thereby opposing the putative negative consensus which is said to have dominated the period prior to the 1950s. Part of his refutation comes in the form of a quotation from the American jazz writer Barry Ulanov who addressed the issue with a dry humour that matches the dryness he detected in this music. He claimed that Boogie-woogie "is much too constricted rhythmically and much too dry harmonically, to have the triumphant aphrodisiac effect upon youngsters which Dr Rodzinsky attributes to it."[36]

These examples of 1940s writing illustrate some of the variety and complexity of the issues raised by the question of how the British imagined America during the period. By including various kinds of non-fictional prose alongside prose fiction I have tried to widen the scope of the enquiry, but this is clearly not intended as a comprehensive survey of writing on this theme. What these texts have in common is that they raise issues which were to resonate into the 1950s and beyond, and the principal value of the discussion lies there.

I would like now to present, by way of triangulation, another approach to the theme of this paper. Having looked so far at "literary" texts, it might be interesting to turn now to some less prestigious sources of evidence. Although the aims of Mass Observation were the subject of disagreement even amongst its originators, there are aspects which at first sight should make it a fruitful source of material in relation to the topic in hand. It had its origins in anthropology, and sought to bring an anthropological eye to the everyday activities and opinions of ordinary people. But its strangely mixed origins also included surrealism, documentary and poetry, and so the alignment with literary texts may not be so difficult. The further connection with the writings of Treece and MacNeice is that they are autobiographical in form, and so can be related to the mass observation material which is based on personal experience and direct observation. In terms of reading and analysing, of especial significance is the emphasis the founders of Mass Observation placed upon the image. There was a belief that from the various reports made by the mass observers a set of collective images would emerge, revealing a deeper level of public opinion than that for example, purveyed in popular journalism

The following example illustrates the way in which an American might imagine that the British constructed a mental image of America. There is a danger of an infinite regress here—what did the Americans think that the British thought that Americans thought and so on ad infinitum. The justification of this approach is that it is important to consider the circulation and re-circulation of "imaginary Americas" to and fro across the Atlantic and not simply their construction and dissemination within Britain.

In 1947 Mass Observation published a report based on surveys of attitudes towards America and Americans. The report commented on some of the adversely critical views expressed by respondents. My purpose is not to address the validity or otherwise of Mass-Observation methodology in the collection of these opinions; the significance of this Mass Observation report in relation to this discussion is that it was the occasion for an American journalist to comment on the way in which, in his view, constructions of America in Britain form a kind of identikit made up of stereotypical images, the result being a grotesque surreal collage:

> Put Betty Grable in a chartreuse bathing suit, seat her in a scarlet 47 convertible about a mile long, hang so many emeralds on her they clank, stuff a cigarette and a wad of gum in her mouth, dangle a huge steak from her belt, put a noggin of rye in one hand and an atom bomb in another, and there you have it. [37]

This bizarre image is a reminder of Mass Observation's roots in surrealism, and could almost be read as a parodic version of the "trait complexes" used to produce a composite group portrait, which was one of the aims of the originators of the Mass Observation project. More generally, we might speculate that weird images of this kind form part of that "everyday surrealism" which is a recurrent feature of the way in which one nation's supposed characteristics figure in the imagination of another. The image-recycling process exemplified in this instance involves a report discussed in the British press, reported in America, then replayed again in a British newspaper. This moderately amusing travesty is an example of the way in which the construction of images of America involves cultural exchange, a circulation of meanings from Britain to

America and back again. There is, to go back to MacNeice's image, a two-way traffic in the "tunnel between the two worlds."

In conclusion, I'd like to end with a sceptical reflection on the possibility of gaining any kind of definitive answer to the overarching question of how America has figured in the British imagination. Again, the Mass Observation archive offers intriguing evidence this time from a survey conducted in March 1945. The examples of various writings which I have already discussed indicate that far from there being a negative consensus, it was possible in the 1940s, outside the discourse of Americanisation, for America to be imagined in varied, complex and often ambiguous ways. British people were also quite capable of questioning some common assumptions. This constitutes a reminder of the danger of indulging in the "condescension of posterity"[38] when generalising about the way in which past generations have imagined other nations. The Mass Observation respondent commented as follows: "To have an opinion about one nation and a different one about another would be assuming the existence of national characters, itself an absolute illusion."[39] In the context of a discussion about the way in which America figured in the British imagination this willingness to question the terms of the debate is salutary.

---

[1] Dick Hebdige, *Hiding in the Light.* (London, 1988.)

[2] D. Webster, *Looka Yonder*. (London, 1988). Webster's book makes an important contribution to the debate initiated by Hebdige, and his observation that positive images of America operated "out of sight of the discourse of Americanisation" encapsulates a significant part of my argument in this paper.

[3] Alan Munton comments that MacNeice's writing is "part of the cultural internationalism that arose in America and Europe during the Second World War." Alan Munton, "Out of Time in America: Louis MacNeice 1939-1943." *Symbiosis*, (April 1997) 132.

[4] Hoggart, quoted in Hebdige, *Hiding in the Light*, 57

[5] Orwell, *Coming Up for Air,* quoted in Hebdige, *Hiding in the Light,* 58

[6] Hebdige, *Hiding in the Light*, 71.

[7] See T Cresswell and B Haskin. "The Kind of Beat Which is Currently Popular. American Popular Music in Britain." In *The American Century*, edited by David Slater and Peter Taylor. (Oxford, 1999).

[8] Hebdige, *Hiding in the Light*, 49

[9] Christopher Brookeman, "The Haunted Wood: Ideology and Culture in the 1930s and 1940s, The Cultural moment of Horizon." *Symbiosis*, (Oct 1998), 178.

[10] Cyril Connolly, "Introduction." *Horizon*, 17.98 (1948), 76.

[11] ibid.

[12] Evelyn Waugh, "The Loved One: An Anglo-American Tragedy." Horizon, 17.98, (1948), 78.

[13] Connolly, "Introduction," 76

[14] Waugh, "The Loved One," 159

---

[15] ibid, 152
[16] Hutnyk, John. *The Critique of Exotica: Music, Politics and the Culture Industry.* London, 2000), 114-15
[17] Olivia Manning (writing as R.D. Smith), *New Statesman*, 11 December 1948.
[18] Christopher Hitchens, *Blood, Class, and Nostalgia.* (London, 1990), 24.
[19] Max Berger, *The British Traveller in America 1836-1860* (1943). (Gloucester (Mass), 1964), 64
[20] Louis MacNeice, "The Way We Live Now IV." *Penguin New Writing*, No 5 (1941), 14.
[21] A view anticipated by Richard Pells: "…much of the conversation between the United States and Europe, before and after 1945, has been characterized more by an exchange of metaphors than by a sharing of information." Richard Pells, *Not Like Us*. (New York, 1997), 2
[22] There is an excellent discussion of this in John Baxendale, and Christopher Pawling. *Narrating the Thirties*. (London, 1996.)
[23] See Munton's persuasive discussion of this change of mood. Munton, "Out of Time"
[24] Munton, "Out of Time," 115-116
[25] Parts of the following analysis have been informed by Amanda Lagerkvist's illuminating article "We See America: Mediated and Mobile Gazes in Swedish post-War Travelogues." *International Journal of Cultural Studies*, Volume 7.3 (2004), 321-342.
[26] Heidi Macpherson, and Will Kaufman. "Transatlantic Literature as Critical Resistance to Americanisation." In *Issues in Americanisation and Culture*, edited by Neil Campbell, Jude Davies and George McKay, 197-210. (Edinburgh, 2004), 199.
[27] MacNeice, "The Way We Live Now IV," 13.
[28] ibid., 14.
[29] Louis MacNeice, "Touching America." *Horizon*, 3.15 (March 1941), 208
[30] Peter Bailey, "Jazz at the Spirella: Coming of Age in Coventry in the 1950s." In *Moments of Modernity*, edited by Becky Conekin, Frank Mort & Chris Waters (London, 1999), 23.
[31] H. Treece, *How I See Apocalypse* (1946) quoted in Tolley, A.T. *The Poetry of the Forties*. (Manchester, 1985), 109.
[32] Bailey, "Jazz at the Spirella" 23
[33] Henry Treece, "Henry Goes Honky Tonk." *Modern Reading*, 13 (1945), 89.
[34] A later variant of this "fictive" approach to America can be seen in Larkin's tongue in cheek reply to a question about whether he had visited America:
"…And of course I'm so deaf now that I shouldn't dare. Someone would say, What about Ashbery, and I'd say, I'd prefer strawberry, that kind of thing. I suppose everyone has his own dream of America. A writer once said to me, If you ever go to America, go either to the East Coast or the West Coast: the rest is a desert full of bigots. That's what I think I'd like: where if you help a girl trim the Christmas tree you're regarded as engaged, and her brothers start oiling their shotguns if you don't call on the minister. A version of pastoral." Philip Larkin, *Required Writing.* London: Faber and Faber, 1983., 70.

---

[35] George McKay, Circular Breathing: The Cultural Politics of Jazz in Britain. (Durham N.C., 2005) 22.
[36] quoted in Treece, "Henry Goes Honky Tonk," 84.
[37] Originally published in the *New York Post*, and reprinted in the British press.
[38] E.P. Thompson's phrase in the Preface to *The Making of the English Working Class* (Harmondsworth, 1963), 12.
[39] Mass Observation, File Report 2222, The Americans (March 1945 Directive).

CHAPTER TWELVE

THE IMAGINING OF AMERICA IN KINGSLEY AMIS'S FICTION

TAMMY GRIMSHAW

Benedict Anderson's groundbreaking study on nationalism suggests that the imagination plays a key role in the formation of communities and nations. According to Anderson, the novel is one of the key tools that communities and nations use to imagine themselves, meaning that communities and nations become constituted, in part, by these novelistic imaginings.[1] But while Anderson discusses how communities use the novel and other forms of written discourse to imagine themselves, Malcolm Bradbury focuses on the way in which writers from various nations use the novel to imagine one another. Explaining that storytelling and mythmaking are two of the primary outlets for the imagination, Bradbury points out that the novel provides fertile ground for this type of vicarious imagining of communities. In his study, Bradbury takes a look at representations of America in the British novel—representations which he claims have grown and developed over the centuries, creating "a common and infinitely expandable imaginative community."[2] Bradbury's assertion has particular resonance when considering Britain's continued political alliance with the United States throughout the twentieth and twenty-first centuries, epochs marked by major wars and international political upheavals. Because commentary on this transatlantic political alliance is often portrayed in the novel, this essay will begin by exploring the imagining of America in British literature against the backdrop of these political events.

Many British writers have depicted, with varying degrees of success, the manner in which Britons come into contact with America and American culture. For some British writers, these depictions were possible without ever having visited the United States. In British literature prior to the nineteenth century, "imaginary Americas prospered quite as well, often better, than did the real one," showing that many British writers did not deem "strict realities or hard facts of history" to be necessary for the creation of their art. Indeed, Spenser,

Shakespeare, Donne, Marvell, Hobbes and Pope all wrote about America in their works without ever having even crossed the Atlantic.[3]

As Max Stites discusses elsewhere in this collection, one of the earliest British novelists to engage in fictional portrayals of Americans after having actually visited the country was Charles Dickens, who incorporated an extended American episode into his novel *The Life and Adventures of Martin Chuzzlewit*, published in book form in 1844 shortly after the author's trip to America in 1842. H. G. Wells, who did not visit the United States until 1905, also made a noteworthy contribution to British portrayals of America—albeit unwittingly—in his futuristic story *The War of the Worlds*, printed in 1898 in an American newspaper that changed the setting of the story from London to Manhattan without the author's permission. In addition, D. H. Lawrence visited the United States in 1922, thereafter writing his highly acclaimed work of non-fiction entitled *Studies in Classic American Literature*, which analyses the struggle for modernity in American fiction of the early twentieth-century.

Even though the aftermath of the First World War meant that most British modernist novelists focused on representing the world closer to home, Aldous Huxley depicted America in a futuristic fantasy, revealing his own distrust of modernity, in *Brave New World* (1932). Malcolm Lowry's lesser-known novel *Delirium on the East River*, first published in 1935 and posthumously re-published in 1968 under the title *Lunar Caustic*, tried to mimic the modernist form in depicting the adventures of Englishman Bill Plantagenet, who seeks artistic and spiritual wholeness in America.

British fictional portrayals of America after the Second World War continued predominantly to deploy the comic form to illustrate clashes of British and American culture. Evelyn Waugh's *The Loved One: An Anglo-American Tragedy*, published in 1948, is a black comedy that challenges America's hope of renewal through British irony. Comic representations of Britons' encounters with American life are also present in Anthony Bailey's *The Mother Tongue* (1963), Julian Mitchell's *As Far As You Can Go* (1963), Pamela Hansford Johnson's *Night and Silence* (1963), Andrew Sinclair's *The Hallelujah Bum* (1963) and Thomas Hinde's *High* (1968). However, perhaps more well known is Malcolm Bradbury's *Stepping Westward* (1965), written after the author's years in America at Indiana University, a moral comedy of manners that depicts the Englishman James Walker's cultural clash with the American version of post-war liberalism, as well as McCarthyism. David Lodge's *Changing Places* (1975), set in 1969 and loosely based on the author's personal experiences in the United States, is also a comedy, albeit with postmodernist elements, that depicts the British protagonist Philip Swallow's encounter with permissiveness and political rebellion at the fictitious Euphoria State University. Moreover, William Boyd's *Stars and Bars* (1984) and Martin

Amis's *Money: A Suicide Note* (1984), written during the decadent decade of Reaganomics, present protagonists whose contact with the abundance of American culture leads them to question their sense of self.

While these works engage in portrayals of Britons' views of America that are humorous by virtue of amusing and playful irreverence, the British novel of the twenty-first century takes on a somewhat different tone. For instance, in *Fury* (2001) Salman Rushdie's protagonist Malik Solanka is forced to confront the anxieties and violence of American urban life at the beginning of the new millennium. And with the attacks on America on September 11, 2001, the tone of these portrayals becomes more extreme. Iain Banks's *Dead Air* (2002) and Martin Amis's *Yellow Dog* (2003) are political satires on the chaos that characterises the post-9/11 world order, while Ian McEwan's *Saturday* (2005) is a diatribe against the United States' "War on Terrorism." Martin Amis takes his rendition of political commentary further in his forthcoming volume of short stories entitled *House of Meetings* (2006), in which he imagines the last days of Mohamed Atta, the infamous lead hijacker of the 9/11 attacks.

One could argue that a direct engagement with political commentary like that in certain post-9/11 British novels is invective in nature, and as such, may help to fuel the already-burning fires of exigent transatlantic political discord. As Andrew Stott points out, political satire is the most "directly political of comic forms." On the other hand, the traditional comic form provides political and social commentary in a much more indirect manner since traditional comedy is driven by the plot, moving it towards a resolution.[4] Because of this indirectness, the treatment of political themes in the traditional comic form is often seen as more light-hearted than that of political satire. While Martin Amis has recently engaged with political satire, his father, Kingsley Amis was renowned for his use of the traditional comic form. Published in 1963, Kingsley Amis's *One Fat Englishman* has been attributed with marking a shift in the tone of the representations of America in British fiction, establishing a tradition of fiction writing that was more highly comic.[5] Amis's comedic portrayals of America in his novels of the 1960s therefore make for an intriguing analysis, particularly given the transatlantic political tensions that existed throughout this decade.

Published shortly after the Cuban Missile Crisis and the initial deployment of Agent Orange in Vietnam, *One Fat Englishman* appeared on the literary scene during an era of intense anti-imperialism and anti-American feeling worldwide. The Vietnam War certainly intensified global feelings of anti-Americanism, and many around the world deplored what they saw as the arrogant American attitude that the United States "was not a classical imperial power, but a righter of wrongs around the world, in pursuit of tyranny, in defence of freedom no matter the place or cost."[6] In addition, the Suez crisis in

1956, when the United States forced Britain to abandon its attack on Egypt, and the activities of the Campaign for Nuclear Disarmament during the 1950s and 1960s heightened anti-Americanism around the world throughout this period.[7] Some would even claim that these anti-American sentiments are alive and well today since the continuing contemporary disclosures about the Vietnam War, as well as the United States' current involvement in the Persian Gulf can, for some, "only be described as imperialist."[8]

Strongly anti-American attitudes were expressed in certain contemporaneous works of British non-fiction during the 1950s and 1960s. These anti-American viewpoints developed because Britain was seen by many as "an accomplice in the USA's postwar neo-imperialism" in Vietnam and other countries by virtue of being political allies with the United States.[9] Besides fearing America's effect on Britain's political system, some also worried about the impact of America on Britain's culture. Several cultural critics, including Leavis, Hoggart and Williams, expressed concerns about the Americanisation of British culture during the postwar period.

Although representations of America in British fiction during the postwar period were predominantly characterised by the use of comedy to make light of the cultural and political differences between the two countries, anti-American sentiments were nevertheless portrayed in certain works of literature written after the Second World War. George Orwells's *Nineteen Eighty-Four* (1949) implicitly criticises the pervasive influence of the American political system on postwar Britain. Airstrip One, the outpost for America's Cold War involvement in Europe, symbolises Great Britain. The currency used in this novel is the dollar, and Americanisms permeate the language.[10] Graham Greene's popular *The Quiet American* (1955), which depicts a young, high-minded American who attempts to channel aid to what he sees as more deserving recipients during the French Army's conflict in Southeast Asia with the Vietminh, may have provoked anti-American feeling amongst its readership. Similarly, *The Ugly American*, first published in 1958, exposed American arrogance, incompetence and corruption in the losing struggle against Communism in Southeast Asia. Written from an engaging fictional first-person perspective, it was widely read, in both the United States and the United Kingdom, and provided much food for thought for those who read it.

Kingsley Amis was greatly troubled by the United States' imperialist policy over Southeast Asia, and in 1967 he "became involved in a long and acrimonious correspondence" about the issues surrounding America's involvement in Vietnam.[11] Given Amis's personal involvement in the political issues surrounding the Vietnam War, critics have commented that *One Fat Englishman* shows a reaction against American imperialism, making this novel "an incisive exploration of the anti-Americanism" of the 1960s.[12] A close reading of *One Fat Englishman*, as well as *I Want It Now*, published in 1968, reveals that Amis used these two novels as a forum in which to explore four particular types of American imperialist attitudes: those about territory, language, the environment and sexuality.

In his portrayal of a cocktail party at the beginning of *One Fat Englishman*, Amis relates anti-American sentiment to America's imperialism when Strode Atkins, an American student, asks Roger Micheldene, the eponymous portly British protagonist:

> Why do you [the British] hate us [Americans]? You do, don't you? You all do. Why? Why? What have we done to you? We didn't want to be world leaders. Last thing we wanted. We've never been imperialist. And yet you hate us. Why? We've never been colonialist. And yet you— (*FE* 45)[13]

Unable to control his rage over these comments, Roger interrupts Strode, snapping back:

> Oh, really? [. . .] Never imperialist or colonialist? What about the Mexican War, the Spanish War? Why do you think places like California and Arizona and Florida and Puerto Rico and the rest them have got those curious foreign-sounding names? (*FE* 45)

Alluding to America's appropriation of territory in previous centuries Roger, a Briton, suggests that Americans have continued to retain an ethos of imperialism, but they simultaneously deny any recollection of the annexation of territory that has made their nation what it is. Hence, this dialogue reveals what Anderson calls the "incompatibility of empire and nation."[14] Just as Britain's colonisation of America was antithetical to the formation of an American nation, America's later appropriation of additional territory worked to quell the identities and communities amongst the indigenous peoples in the territories that America appropriated. Moreover, in portraying Roger's sardonic comments about the origins of certain American place names, Amis shows that the relationship between the "indigenous populations in settled areas and the invading settlers" and "language and the new place" are often inextricably imbricated.[15] The retention of place names in the indigenous language illustrates the way in which the dominator struggles to impose, and is often ineffectual in imposing, its language on those it wishes to dominate in the imperialist conquest.

The struggle over geography or place, according to Said, is essentially a "struggle about ideas, about forms, about images and imaginings."[16] In other words, language is deployed to express images and imaginings about place, creating imagined communities by virtue of the ideas conveyed through language. Yet, it is not merely the ideas conveyed through language that form communities and nations. Rather, language itself helps to form imagined language communities. In this novel, Amis illustrates that language communities are another type of imagined community as he portrays Helene and Ernst Bang, Danish nationals who reside in the United States, disagreeing about their language community membership. Gently chiding his wife for uttering an Americanism, Ernst, a linguistics professor, explains to Helene that "in the Eastern Hemisphere, which as you know includes Scandinavia, the traditional form of English, learnt as a second language, has been British English" (*FE* 14). Scandinavia has adopted British English as a second language because of its geographical proximity to the British Isles, thus becoming part of the imagined community of speakers of British English. Ernst accordingly believes that he and Helene belong to the British English language community by virtue of their nationality. Ernst's comment about the pervasiveness of British English throughout the "Eastern Hemisphere" also reiterates the imperialist conquest since it brings to mind Britain's colonisation of various "non-Western" nations during the nineteenth and twentieth centuries.

In contrast to her husband's assertion, Helene holds the view that she and Ernst are members of the American English language community by virtue of their residence. After Ernst attempts to scold her for using Americanisms, Helene retorts, "But we aren't [living] in the Eastern Hemisphere anymore, we're [living] in America" (*FE* 14). The Bang's young son, Arthur, also believes that, as a resident of America, he is a member of the American English language community. When the British Roger Micheldene challenges a word the Bangs' young son has spelt when the two are playing Scrabble, Roger complains about being given "a bloody American dictionary" to check the spelling of the word. Arthur then retorts, "this is bloody America" (*FE* 60), reiterating the connection between the American English language and his geographical residence or "place" in the United States.

Helene and Arthur's statements demonstrate that the American English language community has emerged as one distinct from its colonisers. That is to say, while the British English language community is associated with the imposition of the English language on others during the imperialist conquest, America did not merely adopt the British English language in order to become part of that imagined language community. Rather, being geographically distant to Britain, the United States has had a history of linguistic struggle and evolution to establish a language that its community members imagine as distinct to its former coloniser's. Ashcroft, Griffiths and Tiffin have commented on this language phenomenon, noting that former settler colonies like the United States have modified and adapted language to allow them to express their sense of uniqueness or difference from their coloniser.[17]

A careful reading of Mr. and Mrs. Bang's conversation reveals that Helene uses Americanised speech to express this sense of uniqueness. Exasperated with her husband, she states, "I still can't figure out why it's so wrong to speak in the American way. More people speak like this than speak in the British way, after all." Ernst then responds, "It's not a question of right and wrong at all. Ideas of correctness don't enter in. Any more than the number of speakers" (*FE* 14). In representing this dialogue, Amis illustrates that the Danish-American Helene is attempting to Americanise her spoken language in order better to become part of her American community. Her husband confirms that language communities can come in all shapes and sizes as he discounts the opinion that the viability of any community is contingent upon the number of its respective members. Moreover, Ernst's comments nullifying the idea of a "correct" form of English bear much resemblance to contemporary theories of language and nationalism. Indeed, Ashcroft, Griffiths and Tiffin have recently asserted that the rise in postcolonial discourses and the diversity of forms of English spoken around the world demonstrate that the belief in a "correct" or "normative" usage of English is illusory.[18]

Attempting to Americanise her speech to achieve the teleological aim of appearing or behaving more like an American, Helene might also modulate her speech with a specific ontological aim. Anderson emphasises that a specific language can offer access to a consideration of one's ontology because it is an inseparable part of that ontology.[19] When considering Anderson's assertion, one could argue that Helene's use of American English emerges as her striving not merely to appear to be American, but in fact to "be" American through her use of language. That Helene's aim is ontological becomes clear as she later comments, "I'm no Dane, damn you, I'm American" (*FE* 46).

Although Amis depicts Professor Bang's belief that there is no such thing as a "correct" form of English, he portrays the British Roger holding a view of English that is diametrically opposed to the Professor's. Amis comically illustrates the way in which ideas about the "correctness" of the English language can create cross-cultural conflict between Britons and Americans in a dialogue between Roger and Irving Macher, an American academic. Lashing out at what he sees as Macher's limited and confined experience of the British English language, Roger says, "Do you think, Mr. Macher, [. . .] that [. . .] you're entitled to pontificate about [. . .] what was that bit of jargon? Forms of speech?" (*FE* 84). Macher laughs a bit longer than necessary and then replies, "As a native American in full possession of his faculties I can claim complete parity with you as a user of English" (*FE* 85).

In this fictional conversation, Roger clings to a view of British English that is decidedly colonial. Anderson discusses the colonial deployment of language at length. He points out that languages are often viewed as "the personal property of quite specific groups" or communities and adds that former colonial powers like Great Britain demanded "isomorphism between the stretch of the various empires and that of their vernaculars."[20] In his acerbic comments to Macher, Roger alludes to the idea that English is the "property" of Britain; therefore, American English cannot be considered a "correct" form of English. This depiction is in keeping with the view that the colonial overlord often "installs a 'standard' version of the [. . .] language as the norm, and marginalizes all 'variants' as impurities."[21] In the same way, Roger sees American English as being an incorrect variant of British English, although Macher asserts that American English is a unique language in its own right, rather than a variant or derivative language.

However, Roger and Macher's dispute is not merely about language use. Their disagreement also concerns the dynamics of power between communities or nations since "language becomes the medium through which a hierarchical structure of power is perpetuated."[22] Amis illustrates the dynamics of power and language by showing that Roger's discomfort with American English springs from this character's frustration over the predominant position that America has come to occupy in world affairs. Ashcroft, Griffiths and Tiffin note that, like the other colonial powers of the nineteenth century, Britain has been "superseded by the emergent power of the USA," and as a result, "has been relegated to a relatively minor place in international affairs." The USA has therefore moved "from a dominated to a dominating position."[23] Amis comically represents these power dynamics in Macher's parting comment to Roger: "As regards the development of the language the U.S.A. is now central and England peripheral" (*FE* 84-85). Here, Amis uses Macher's comment to show that the development of the English language is a metaphor for the quest for world domination, with the United States and Great Britain being the two main contenders in the struggle for power. Importantly, Macher's claim about the centrality of American English worldwide also alludes to the notion that while language communities may be one type of imagined community, domination of others may work to create communities of power that can be reified. Indeed, in spite of the proliferation of recent theories on the imagined nature of communities, Davies finds that "nations [. . .] are not just imaginary states but actual formations of power."[24]

An earlier conversation amongst Roger Micheldene, Irving Macher and Nigel Pargeter, a British student who is doing postgraduate study in the United States, can also be interpreted as Amis's allusion to these national formations of power. Roger tells Pargeter:

> I suppose you think that just because there aren't any dukes here everybody's all chums together. Complete illusion. Look, when the Queen and Prince Philip

> were here and drove through New York or Washington or one of these places they had more of a reception than that frightful man General MacArthur. (*FE* 44)

Wishing to correct Roger, and thereby embarrass him, Macher then interrupts the conversation by referring to the volume of ticker-tape thrown at parades: "I'm sorry, Mr Micheldene, but [. . .] your Queen and Prince rated between one-quarter and one-fifth as much [ticker-tape thrown during their parade] as that (I agree) frightful man MacArthur" (*FE* 44). In his juxtaposition of MacArthur, an American war hero, to the British Royal family, Amis provides an interesting contrast between the different types of imperialism that have historically been present in the United States and the United Kingdom. MacArthur, having spent his career in the United States army and earning the Congressional Medal of Honor, was often considered to be a consummate patriot, and by mentioning MacArthur, Amis connects America to imperialism through military conquest. The Queen and Prince Phillip, of course, relate to imperialism through the direct exercise of colonial power. The passage thus speaks to the imperial contest and concomitant battle between the United States and Britain for empire, or world power and domination.

In a later conversation between Ernst Bang and Roger Micheldene, Amis illustrates how language use also reveals the tendencies behind each country's imperialist impulses. Amis again portrays the connection between language and imperialism as Ernst says to Roger:

> Where an American typically says *I do this,* showing concern with an activity related to an object that is immediately present, an Englishman typically says *I have that,* showing concern with a condition that need be none of his making and is related to an object that may be at a distance in the past. Americans pursue the dollar; the British had an empire. Fascinating to see the underlying assumptions and goals of a culture laid bare in its idiom. (*FE* 146-47)

The idiomatic linguistic features that Ernst describes reveal how imperialist attitudes are expressed through the medium of language. In this dialogue, American behaviour is aligned with consumption and consumerism because of the emphasis on material "objects that [. . . are] immediately present." On the other hand, British culture is associated distancing one's self from one's actions since it relates to past actions for which one does not claim direct responsibility. America's imperialism is therefore often driven by its pursuit of materialism and consumerism. Attributed with distancing one's self from one's actions, Britain's colonialist tendencies may be shrouded in a failure to accept culpability for the deleterious effects of the colonisation of other territories.

One such deleterious effect of the imperialist contest is that of environmental damage to colonised places or territories. Just as Amis depicts the manner in which Americans and Britons have disputes about language, so too does he illustrate the disagreements the two nations have about how best to relate to the natural environment. Being a Briton, Roger in *One Fat Englishman* fails to understand the way in which Americans relate to their natural world. He muses, "He was not clear in his mind how he wanted these people to regard the fauna of their country" (*FE* 52). Amis thereby illustrates how nationality can affect one's interaction with the environment. As Gerrard comments, because place is so important to environmental issues, it is not surprising that America, particularly in its interaction with nature, "has followed its own distinct trajectory as a response to an environmental and social history very different from that of Britain."[25]

In spite of these cross-cultural differences and misunderstandings about nature, it is clear in Roger's mind that certain of Americans' attitudes towards the natural world are annoying. Amis goes on to write that Roger "could have done with less of their [Americans' . . .] excited wonder at harbouring so many species within their borders" (*FE* 52). The sense of amazement and wonder that Amis depicts in this passage demonstrates what Garrard describes as a sublime relationship with nature, a state that is marked by astonishment.[26] While one might expect such a sublime relationship with nature to have a positive impact upon the environment, Roger suggests that a nation's relationship with the natural environment may inhere in attitudes of appropriation that are integrally linked to imperialism, and it is these imperialist attitudes that cause his annoyance.

Roger adds, "It stood to reason that any fool who owned half a continent was going to own a lot of birds and mammals and such as well" (*FE* 52). Roger's comment implies that America's "ownership" or appropriation of the natural environment is a necessary prerequisite to the nation's formation of a relationship with nature and parallels what Alfred Crosby has called ecological imperialism, a phrase used to describe the effects of imperialism on the natural world.[27] However, Amis seems to illustrate that ecological imperialism is inversely related to the experience of the sublime: given that a greater amount of land "owned" leads to a wider diversity of natural life, individuals should experience a heightened or intensified encounter with the sublime as they encounter the increased variety of natural life in the appropriated territories; nevertheless, such excessive appropriation may paradoxically reduce the experience of the sublime and lead to complacency about the environment.

Amis does not portray America completely annihilating the environment, a process that his son Martin has called the "toiletisation of the planet."[28] In spite of this, though, Amis does appear to express concern about Americans' relationships with their natural world by creating a narrative shift later in the novel that might serve as an admonition about being complacent towards the environment. As Garrard points out, interactions with nature may be portrayed as "rhapsodic celebrations," like those of the sublime that Amis depicts in the "excited wonder" mentioned in the previous excerpt, but they may also be jeremiads.[29] Amis's narrative shift is evident as Roger later states: "Nobody was interested in having flowers just growing round the place: who would bother to plant and tend a rose-bed when he could have a Cadillac delivered in an hour?" (*FE* 102). In posing this rhetorical question, Roger claims that Americans may like to interact with the natural environment by observing wildlife, but they certainly do not wish actively to tend and care for their natural world, through gardening, for example. While one must be careful not to elide the voices of narrator and author, these remarks could be read as Amis's oblique warning that paying homage to the acquisition of technological goods, such as petrol-guzzling automobiles, at the cost of giving due care and attention to the environment, can only result in serious ecological harm.

The mention of the Cadillac in this excerpt is of special interest when considering the way in which a nation interacts with its natural world because the comparison of the garden to the Cadillac illustrates the contrast of the rural to the urban that Raymond Williams insisted was ever-present in representations of nature in literature.[30] Yet, a nation's involvement with technology is essentially a clash not only of the rural and urban, but also of the natural and the cultural since "nature" is always in some way culturally constructed.[31] Here, the Cadillac represents the urban and the technological, as well as the imposition of technology on both culture and nature. Recalling the intrusion of the train in Thoreau's woodland paradise, the representation of the Cadillac also brings to mind the novels of Dickens, whose fictional portrayals of American showed that even though "the idealized image of America was the New Eden, the reality was the rattling machine of the railway train, steaming over the prairie and laying waste to all in its path."[32]

But while the Cadillac represents the technological, the cultural and the urban, the tending of the garden exemplifies the natural and the rural. That a garden must be tended to and cared for according to the socio-cultural "norms" that would compel these actions supports the critical viewpoint that nature is culturally constructed. Since Americans are not interested in investing the time and effort to tend to nature, in Roger's mind, the process of cultural construction of the "natural" has failed in this particular case, however.

Dominic Head has recently remarked that "the Green movement in general is predicated on a [. . .] deprivileging of the human subject."[33] Conversely, Roger seems to suggest that Americans are somewhat loath to forgo their privileged position as technologically-reliant human subjects—a reliance that inevitably not only damages the environment, but also creates impediments for the Green movement. If one were to assume that the persona of Roger Micheldene is a guise for the novelist's own fears for environment, Amis might also be taken as implying in this narrative excerpt that Americans are sometimes too self-centred to see the way in which their dependence on technology causes unnecessary ecological harm.

In addition to illustrating the connection between imperialist attitudes and the environment, Amis depicts the male's control over the female as sexual "Other" in certain of his novels as a further metaphor for imperialism or appropriation. These representations of the patriarchal control of women result in depictions of what I would like to call sexual imperialism. The sexual oppression of women parallels processes of imperialism in many ways, and the similarities between feminism and postcolonialism have been well documented in recent research.[34] Women, like colonised subjects, have encountered various forms of patriarchal domination, and they share this experience of oppression with colonised races and cultures. The similarities between imperialism and the oppression of women are also represented in contemporary fiction. Toni Morrison, Paule Marshall and Margaret Atwood have all drawn analogies between the relationships of men and women and those of the imperial power and the colony.[35]

Likewise, Kingsley Amis's portrayals of "masculine" appropriation of "feminine" sexuality bear a striking resemblance to imperialist attitudes. When a female friend of the Bangs persists in pursuing him, Roger displays sexual imperialism as he muses: "American women seemed entirely without finesse. He preferred frank submission to frank pursuit except, theoretically, from the kind of woman who frankly made no move of any kind in his direction" (*FE* 136). In these reflections, Roger reveals not only that he prefers to "conquer" the opposite sex, but also that he likes the females he "conquers" to be passive, making his acquisition or appropriation of them as sexual "Other" a less onerous task. Tellingly, the attractive, self-assured type of woman who might make the first move sexually has no interest in doing so towards Roger, perhaps because she senses his sexually imperialist attitudes.

Roger also makes derisive remarks about what he sees as the aspects of "feminine" sexuality that are unique to American women as he later muses that Grace Derlanger, Macher's girlfriend, is "a Yank bag" (*FE* 23). These musings show that being a "bag" is bad enough in Roger's mind, but being an American one is infinitely worse. He then imputes promiscuity to American women, particularly to the younger generation. Revealing more about his own fantasies than American women's tendencies, Roger muses as one of the female students glances at him, "these Yank college girls were at it all the time, one heard" (*FE* 81).

But although Roger himself considers American women to be sexual "Other," he becomes enraged when he witnesses American men behaving in a sexually imperialist fashion. Upon discovering at the end of the novel that

Macher, his romantic rival, is having a sexual liaison with Helene Bang, whom he wished to "conquer" himself, an enraged Roger tells Macher, "You want something so you just take it. [. . .] Just come crashing in [. . .] . . . Like all your bloody countrymen. It'd be funny if it weren't so terrifying" (*FE* 162).

As Roger accuses Macher of "just taking" Helene to suit his needs, he unwittingly reveals that, as an Englishman, he finds distasteful the manner in which Americans use such appropriation to achieve their sexual aims. And even though Macher's aim in this case is that of a sexual conquest, Roger's accusation recalls his earlier diatribe against Americans' proclivities towards appropriation during his rant on America's annexation of territory in its movement towards westward expansion. Roger then goes on to attribute the penchant for appropriation not only to Macher, but to all Americans generally as he states that Macher is "like all [. . . his] bloody countrymen." Finally, Roger's comment that such behavior is "terrifying" shows that Britons sometimes treat Americans as an "Other" who is terrifying perhaps because America's denial of the harm inflicted upon indigenous peoples in its westward expansion and activities in Vietnam brings to mind the harmful impulse towards appropriation that has also existed in British culture.

Amis continues to portray sexual imperialism in a later novel entitled, *I Want It Now*, published in 1968.[36] Ronnie Appleyard, the British male protagonist in this work, attempts to conquer Simona, a female character who could be classified as half-American, since her mother is an American. Amis portrays Ronnie's conquering of Simona in a way that is similar to Roger's conquest of Helene Bang in *One Fat Englishman*. However, while Roger desires Helene only to satisfy his sexual aims, Ronnie's desire for Simona in *I Want It Now* is not only sexual, but also monetary: he wants her money as well as her body. Because he is worried about losing access to her money, Ronnie later tempers his pique against American "femininity," deciding against "telling her [Simona] to try not to be so crappily American" (*WN* 82).

I Want It Now also demonstrates that Amis's imagining of America was enmeshed with his stance on morality. Often considered a moralist, Amis displayed in his fiction of the 1960s in particular a preoccupation with the "morals and follies of [. . .] sex."[37] The focus on sexuality and morality is clear as Ronnie is arrested by local law enforcement officers in South Carolina after crossing the state line with Simona an arrest that takes place in spite of Simona being over the legal age of consent (*WN* 169). Moseley notes that in *I Want It Now*, Amis "revisits the American scene (this time with a determined attack on American, particularly Southern, traits)."[38] The portrayal of the bumbling and uncouth Mr. Fields in Ronnie's arrest scene provides an indication of Amis's contempt for certain Southern attitudes. Fields states:

> Mr. Appleyard is liable to arrest [. . . for] making [. . .] a false declaration [. . .] in the furtherance of immorality. In addition, Mr. Appleyard has committed a Federal offence under the Mann Act, prohibiting the conveyance of a female over a State border for an immoral purpose. In fact, Appleyard, [. . .] you can take my word for it that we have you hog-tied real nice. (*WN* 169)

In this passage, Amis derides what he saw as the hypocritical Southern tendency to moralise about sex and sexual relationships. In addition, the word "conveyance" recalls the legal transfer of property from one individual to another and reveals that Southern laws dealing with women treat them in an imperialist fashion as property or territory which the most powerful male can claim. Amis's use of tone and register in this excerpt is also significant. The use of the Southern colloquialism "hog-tied real nice" is provided as a coda to this derisive commentary, depicting Americans from the South as boorish not only for using metaphors that compare humans to animals, but also for having grammatical inaccuracies in their vernacular, in this case using the adjective "real" instead of the adverb "really."

However, while Southerners might wish to appear to have a staunch sense of moral rectitude about sex and sexuality, Amis felt that it was highly paradoxical that many from the American South felt no compunction whatsoever about treating their African-American citizens in a morally reprehensible way. Because there is always a "connection between imperial politics and [. . .] hierarchies of race," Amis's concern with racism of course reiterates his quarrel with imperialism.[39] Amis's representation of sexual imperialism or gender oppression in this novel might therefore be enmeshed with the outrage he felt over racist conditions in the American South, meaning that his portrayals of the oppression of women in this novel could be interpreted as a metaphor for racial oppression.

Amis personally witnessed the racism that was prevalent in the Deep South during his stay at Vanderbilt from 1967 to 1968, a time he described in his *Memoirs* as "second only to my army service as the [. . . period] in my life I would least like to relive" and an experience that led him to the conclusion that many people living in this region of America were socially backward.[40] And while racism is always an important issue, it was perhaps even more so when *Want It Now* was published in 1968, months after the assassination of Dr. Martin Luther King. 1968 was also the year of the Tet Offensive, a wave of assaults in Vietnam named after the Asian Tet holiday during which they took place. Resulting in a massive loss of life for the United States troops, the Tet Offensive brought about a marked negative turning point in public opinion over the Asian war effort.[41] Thus, Amis ostensibly used *I Want It Now*, as he did *One Fat Englishman*, as a forum in which to explore imperialist attitudes in general, whether they be directed against women or those of other races or cultures.

So, one lingering question remains: Can one conclude that Amis was anti-American and incited anti-American attitudes because of his sometimes unflattering fictional portrayals of American life? Said has observed that "novels participate in, are part of, contribute to an extremely slow, infinitesimal politics that [. . .] reinforces [. . .] perceptions and attitudes about England and the world."[42] Because of this interplay between literature and society, one could argue that Amis's sometimes negative portrayals of Americans might perpetuate anti-American sentiment. Although I am loath to tread upon the slippery slope of author intention, an examination of Amis's *Memoirs* reveals that some of his sentiments do seem to be anti-American. For instance, he states, "When people try to ape those of another country, do they always admire and emulate the worst parts of the foreign culture? Answer, of course not [. . .]. But in America, sadly, quite often so."[43] It has also been noted that *One Fat Englishman* exhibits "a strident anti-Americanism"—"a view Amis was eager to contest at the time, and indeed [. . . went] on contesting, although [. . . during the 1970s] with rather less force" than he did previously.[44]

Amis spoke out against American literature in particular, claiming that "Americans [. . .] don't actually know anything and are no good at anything, apart from science. They're no good at literature. [. . .] I think that most American literature is a disaster." Yet, Amis also remarks that there is "a sad rift between British and American literature which has done so much to impede our common cultural understanding."[45] One can only note that it is puzzling that Amis decries transatlantic misunderstandings, while perhaps perpetuating these very misunderstandings not only in his denunciation of American literature, but also in the representation of America and Americans in his fiction. It could even be argued that Amis's fulminations against American literature were imperialist in and of themselves because, in making these judgments, he falls prey to the tendency to judge American literature with Anglo-centric attitudes.[46]

Militating against such a narrow interpretation, however, one must note that Amis describes himself in his autobiography as: "strongly pro-American in my attitudes." Amis also talks about his experience as a visiting fellow in creative writing at Princeton from 1958 to 1959 in terms that are generally positive, even to the point of commenting that the United States "was my second country and always would be."[47] It could be said that Amis had a balanced view of American life, seeing both its positive and negative characteristics. He describes the "general picture" of the United States variously, stating that America exemplifies: "energy, generosity, goodwill [. . .], affluence, [. . .] capitalism, violence, squalor and ugliness, [. . . and] beauty."[48]

In light of Amis's rather contradictory views on America in general and American literature in particular, critics have concluded that the tendency to claim that Amis is anti-American stems specifically from two key factors. First was his contempt of the manner in which American writers often held down university posts in order to have secure incomes. He believed that university posts stifled the American writer's creativity, and his poor opinion of American literature was a direct result of this belief. That is to say, Amis derided the manner in which American writers had "the academic cushion"—university posts whose income supported their work as novelists—a phenomenon that he believed detracted from their sustained and focused attention on their work as writers.[49] The second factor to consider is the tendency of readers naively to view Amis's fiction as covert autobiography.[50] Amis is often mistakenly identified as the autobiographical or confessional voice of his male protagonists. Yet, as Moseley points out, the anti-Americanism in *One Fat Englishman* is Roger Micheldene's, not Amis's, and "Amis has carefully constructed Roger's character so as to deny him 'authority'."[51]

To my mind, reading Amis's fiction as anti-American because of the seemingly anti-American statements he made in his memoirs would be much too reductive and runs the risk of falling prey to the intentional fallacy. Importantly, critics have noted that the anti-Americanism in Amis's fiction is thoroughly tongue-in-cheek since his seemingly negative portrayals of Americans are tempered by the use of comedy. In an interview, Amis expounded on his use of comedy in characterisations like Roger's, stating that "it's essential from my point of view that the bad people [i.e., characters in his novels] should be ridiculous as well as bad."[52]

Diffusing the effect and impact of the seemingly anti-American sentiments throughout his fiction through the use of comedy and irony, Amis equally makes fun of the way that Americans imagine and communicate with the British, as well as using comedy to poke fun at his own nationality. That Amis's depictions of cross-cultural misunderstandings are comic is most apparent in *One Fat Englishman*. For example, in "The Game" the Bangs play at dinner parties, guests have to act out an adverb in the form of charades. In a highly comic scene, Amis portrays Roger taking umbrage when Helene asks him to enact the adverb "Britishly" (*FE* 40). Amis's use of comedy gently to ridicule the peccadilloes of his own nationality is also evident later as Professor Castlemaine notices that Roger is rather upset when he is searching for Irving Macher, his rival for Helene Bang's extramarital affections. Castlemaine then says, "Mr. Micheldene, I do believe you're agitated. I thought the British were never agitated" (*FE* 149).

One of Amis's most tongue-in-cheek passages at the end of *One Fat Englishman* perhaps best sums up the author's use of the comic form to portray not only cross-cultural, but also interpersonal, conflict as the American Irving Macher makes this parting comment to Roger in New York:

> I sometimes got the impression that you think some of the people in this country don't like you because you're British. That isn't so. We're out of the redcoat era now, even if you aren't. And we don't think this way. We don't have group likes and dislikes. It isn't your nationality we don't like, it's you. (*FE* 168)

---

[1] Benedict Anderson, *Imagined Communities: Reflections on the Origin and Spread of Nationalism*, rev. edn. (London, 1991), 144. For a further analysis of the impact of language and discourse on nationalism, see Homi K. Bhabha, *The Location of Culture* (London, 1994).

[2] Malcolm Bradbury, *Dangerous Pilgrimages: Trans-Atlantic Mythologies and the Novel* (London, 1995), 485.

[3] Ibid., 6.

[4] Andrew Stott, *Comedy* (New York, 2005), 109, 41.

[5] Bradbury, 449.

[6] Edward Said, *Culture and Imperialism* (London, 1993), 3.

[7] Alistair Davies, "Britain, Europe and Americanisation," in *British Culture of the Postwar: An Introduction to Literature and Society 1945-1999*, ed. Alistair Davies and Alan Sinfield (London, 2000), 103-09 (105).

[8] Said, 64.

[9] Davies, "Britain, Europe and Americanisation," 106. Davies also points out that television and cinema during the Cold War period portrayed the American way of life as an overwhelmingly positive experience, in contrast to the sometimes negative portrayals of Americans in British fiction and non-fiction during this era (106).

[10] Ibid., 104.

[11] John McDermott, *Kingsley Amis: An English Moralist* (Houndsmills, 1989), 3.

[12] Merritt Moseley, *Understanding Kingsley Amis* (Columbia, South Carolina, 1993), 2, 65.

[13] Kingsley Amis, *One Fat Englishman* (Harmondsworth, 1966). This novel will be referred to in the text with the abbreviation *FE*.

[14] Anderson, 93.

[15] Bill Ashcroft, Gareth Griffiths and Helen Tiffin, *The Empire Writes Back: Theory and Practice in Post-colonial Literatures*, 2nd edn. (London, 1989), 133.

[16] Said, 6.

[17] Ashcroft, Griffiths and Tiffin, 11, 7.

[18] Ibid., 37.

[19] Anderson, 36.

[20] Anderson, 84, 77; Said describes this situation as "the predicament of sharing a language with the colonial overlord." See Said, 274.

[21] Ashcroft, Griffiths and Tiffin, 11, 7.

[22] Ibid., 7.

[23] Ibid., 6-7, 31.

[24] Alistair Davies, "From Imperial to Post-imperial Britain," in *British Culture of the Postwar: An Introduction to Literature and Society 1945-1999*, ed. Alistair Davies and Alan Sinfield (London, 2000), 1-8 (4).

[25] Greg Garrard, *Ecocriticism* (London, 2004), 34.

[26] Ibid., 64.
[27] See Alfred Crosby, *Ecological Imperialism: The Biological Expansion of Europe, 900—1900* (Cambridge, 1986).
[28] See Cynthia Deitering, "The Postnatural Novel: Toxic Consciousness in Fiction of the 1980s," in *The Ecocriticism Reader: Landmarks in Literary Ecology*, ed. Cheryll Glotfelty and Harold Fromm (Athens, Georgia, 1996), 196-203 (196).
[29] Garrard, 81.
[30] Raymond Williams, *The City and the Country* (London, 1973), 165.
[31] Garrard, 9-10.
[32] Bradbury, 274.
[33] Dominic Head, "The (Im)possibility of Ecocriticism," in *Writing the Environment: Ecocriticism and Literature*, ed. Richard Kerridge and Neil Sammells (London, 1998), 27-39 (28).
[34] For example, see Gayatri Chakravorty Spivak, "Can the Cubaltern Speak?" in *Marxism and the Interpretation of Culture*, ed. Cary Nelson and Lawrence Grossberg (London, 1988).
[35] Ashcroft, Griffiths and Tiffin, 30.
[36] Kingsley Amis, *I Want It Now* (St. Albans, 1970). This novel will be referred to in the text with the abbreviation *WN*.
[37] Moseley, 4.
[38] Ibid., 79-80.
[39] Said, 7.
[40] Kingsley Amis, *Memoirs* (London, 1991), 282-94.
[41] See Moseley, 80.
[42] Said, 89.
[43] Amis, *Memoirs*, 205.
[44] McDermott, 119.
[45] Amis, *Memoirs*, 197.
[46] For more information on these Anglo-centric attitudes, see Ashcroft, Griffiths and Tiffin, 4
[47] Amis, *Memoirs*, 193-94.
[48] Ibid., 211.
[49] Moseley, 8-9.
[50] McDermott, 2.
[51] Moseley, 66-68.
[52] Dale Salwak, "An Interview with Kingsley Amis," in *Interviews with Contemporary Writers: Second Series, 1972—1982*, ed. L. S. Dembo (Madison, Wisconsin, 1983), 112-29 (116).

CHAPTER THIRTEEN

THE BUFFALO (THEATRE) BILL: SELLING SAM SHEPARD'S AMERICAN VISION IN 1970S LONDON

TIMOTHY ROGERS

Introduction

The artistic career of Off-Off-Broadway playwright Sam Shepard has spanned over four-decades and fifty plays. He also has been identified as musician, novelist, director, and perhaps most visibly, Hollywood star, appearing in over sixty films. Shepard's many facets and his professional longevity have left some areas of his work open to deeper scrutiny due to a particular study's breadth of scope. One area of interest has been his self-exile from America, between 1971-1974, when he and his wife, actress and composer O-Lan Jones, took their newborn son Jesse to reside primarily in London. During this three-year period, Shepard produced two of his best-known pieces, *The Tooth of Crime* and *Action*, along with *Geography of a Horse Dreamer* and the more obscure *Blue Bitch* and *Little Ocean*.

This paper looks to broaden the scope of Steven Putzel's argument that the plays of Shepard's London years were met with patronising condescension from the London press and interpreted as a sort of savage exotica reminiscent of Buffalo Bill's Wild West show.[1] Such a critical response to American culture as both savage and at the same time completely disassociated from British culture reinforces a feeling of smugness that sidetracks a confrontation between a British audience and elements of savagery in British culture:

> …the London audience experiences no self-examination and painful self-knowledge, but instead revels in an evening of American exoticism and campiness that comfortably confirms its view of the wild ex-colony festering a safe three thousand miles away; this frontier society both repels and provokes 'giggles' from the superior British.[2]

Putzel uses the term "horizon of expectations" and its emphasis on the implied audience over authorial intentions to suggest that these plays, transplanted as they were from their American soil, would have been approached by a different set of cultural values other than those that informed their writing. In having looked at a number of promotional materials from the Royal Court archives, I would add that beyond the cross-cultural transplant of the texts, the publicity of their productions would have been a contributing factor in the reinforcement of (and occasional challenge to) this horizon, in addition to myriad artistic and technical choices made by those realizing the texts in performance.

In this sense I shall be following Carol Benet's example in *The German Reception of Sam Shepard* (1993) which considers a pool of information often displaced in an incompatible cultural context as contributing to the hostile reception of Shepard's work in Germany. For Benet, where there was initially little information about the playwright or his plays whatsoever (as was very much the case in 1970s London), publicity by the host theatre company played a more important role in setting this horizon of expectations than that of an already established playwright. Despite the potential for common linguistic ground between American and British use of language, I would argue that Shepard's plays may have been the site of British misunderstanding and projection of anxieties not unlike Shepard's German reception.

Whilst I shall only focus on publicity materials from the Royal Court, it is true Shepard worked with the Open Space and the Hampstead Theatre Club (as well as television studios), and no doubt their publicity materials were another factor yet to be considered in a circular transmission of information drawing from and feeding back into itself. It is beyond the scope of this paper, moreover, to recreate a sort of "big bang" where this British horizon of expectations formed, if this is even possible to speculate. For there had been a very limited number of Shepard productions prior to those this paper addresses, which would not have received as much press attention, yet which clearly cannot be dismissed despite the lack of archival evidence. Further, the input of Shepard's former agent, Toby Cole, who was also the American agent for Court playwrights such as Edward Bond, would warrant a wider future study. It is also impossible to prove how much control Shepard had over publicity material, though greater control is evinced in the third of the plays to be considered.

Whilst it might be impossible to prove any deliberate self-promotional strategy on Shepard's part, certainly aspects of his persona played into an image of the American primitive that the London press responded to in their reception of his plays. In fact, it may be easier to argue that Shepard was not interested in promoting himself as a playwright except out of financial necessity, for compared with earlier periods when he often produced five to eight new plays a year, this less prolific literary output (of five finished plays over three years) had been precipitated by a near two-year hiatus from writing when Shepard pursued musical interests, which culminated with Shepard's English sojourn to "…somehow fall into a rock n' roll band. It didn't work…."[3]

Though Shepard was a skilled drummer with the Moray Eels, (a.k.a. the Holy Modal Rounders), by some accounts it seems clear why his hopes to break into rock n' roll were ill-founded. The monotone nature of the music for *The Tooth of Crime* was dismissed by critics, and later by Shepard himself.[4] In a recent e-mail, playwright and actor Michael Weller, a fellow expatriate then in London, discussed Shepard's inability to relay simple musical directions for the Open Space 1972 world premiere of *Tooth*, "He tried to show us the chords to the songs he wrote, but it was hard to follow… he's not a trained musician." This was something Shepard likewise suffered when working with the band Blunderpuss who accompanied these songs.[5]

Shepard's failure to enter the music industry left him struggling to support his family in a country crippled by mounting inflation, massive unemployment, three- and two-day weeks and union industrial action. Shepard's financial woes "living off his meagre royalties and grants in Hampstead" were shared by many artists at the time. Another expatriate, Nancy Meckler, director of *Action*, recounted originally meeting Shepard when O-Lan joined Meckler's Freehold Theatre, an offshoot of Ellen Stewart's La Mama troupe in New York. O-Lan was unable to find work, Meckler said, and hoped taking acting workshops might prove professionally lucrative.[6] "None of us had any money then…it was a very difficult time to find any work at all, especially in theatre."[7]

Thus, despite Shepard's move away from theatre, he found himself writing plays to survive. It is perhaps not so ironic that during this time he also made an institutional shift from the relaxed atmosphere of 1960s Village counter-culture towards more conservative organizations such as the BBC, Granada Television and the English Stage Company at the Royal Court. The Court especially proved a major proponent of Shepard, where both *Geography of a Horse Dreamer* and *Action* had their world premieres in 1974. The Court also staged revivals of *La Turista* (not long before he arrived in 1969) *The Unseen Hand* (1973) and *The Tooth of Crime* (1974, less than two years after its world premiere at the Open Space). Such alignments were likely opportunistic on Shepard's part, with the offer of funding, such as the Court's award of its Neville Blond Fund to Shepard for *Geography of a Horse Dreamer*.

More ironic still was Shepard's acceptance into the world of London theatre, perhaps abetted by some spuriously accidental transatlantic publicity of Shepard's prior, more high-profile work, despite the fact that much of this had proven professionally disastrous for him. These included the prestigious Lincoln Centre premiere of his *Operation Sidewinder* in 1970, which received almost unanimously poor reviews, save that of Catherine Hughes which appeared in *Plays and Players*, and his work on Michelangelo Antonioni's *Zabriskie Point*, (which Shepard disdained and from whose screenwriting credits he tried to have his name removed).[8] Further, according to producer Albert Poland, at around this time Shepard was trying to secure a place for his work in the New York theatrical mainstream as it was being eulogised by its more traditional critics. Indicative of this was Clive Barnes of the *New York Times*, who infamously analogised Shepard's "disposable" *Operation Sidewinder* on 2 April 1970 as the equivalent of "what Kleenex was to the handkerchief."[9]

Despite this possibly well-manoeuvred aura of literary promise, London theatre was not initially primed to welcome Shepard with open arms, even amongst those familiar with his work. Another American expatriate, Charles Marowitz, whose Open Space premiered *The Tooth of Crime*, was loathe to take on what he called the "Shepard hard-sell which had been emanating from the States" when Shepard approached him to stage the piece.[10] Further, prior to 1971, apart from a few whirlwind readings and touring productions by the La Mama troupe in 1967, the Court's revival of *La Turista* at their newly opened Theatre Upstairs in 1969 and the 1972 Open Space premiere of *the Tooth of Crime*, Shepard had received little press coverage.

If anything, however, London theatre welcomed Shepard for his exoticism. "They were delighted by his drawling, down-to-earth James Dean persona," and Shepard, with characteristic opportunism, appeared to encourage such perceptions. Even Marowitz, despite eventual antagonisms that would lead to a lifelong breach with Shepard, admired his "conquering charm that is sometimes bred in the southern and western sections of America…he remains Huck Finn minus the fishing rod." Former Court Artistic Director Nicholas Wright claimed in a telephone interview that "Shepard was far more exotic than Wole Soyinka" whilst in another telephone interview, Peter Gill, heavily affiliated with the Court in the early 1970s, playfully recounted Shepard as "a six-foot tall Stetson hat wandering around Hampton Heath."[11]

The Court's intrigue with Shepard's embodiment of American exoticism reflects an atmosphere that Max Stafford-Clark, who was just beginning at the Court when Shepard was there, recalled in a telephone interview. "If there's anything about the Court that speaks of a 'Court style', it would be its relentless penchant for bringing other worlds to audiences."[12] Stafford-Clark was referring to, amongst other things, former Artistic Director Oscar Lewenstein's attempt to stage as many foreign as British playwrights despite fierce opposition from Court conservatives such as Lindsay Anderson.[13] Lewenstein's appointment to this post was informed by the Court's need to compete in the commercial sector of London theatre, and the veteran impresario's sense of what would sell no doubt was seen as an asset even amongst more reactive Court conservatives.

Further, Shepard's trademark aura of mystique invited assumptions and projections onto his rugged, taciturn image that made him all the more exotic to the Court. Even in 1974, when Shepard was firmly entrenched in the London arts scene, there was a lack of information about the playwright. Nicholas de Jongh, in reviewing a lunch-time reading of *Cowboy Mouth* for the *Guardian*, 22 March 1974, cited it as co-written by Shepard and rock star Patti Smith (which in itself is correct) but incorrectly referred to Smith as Shepard's wife, though the two were having a much-publicised affair at the time of the play's writing. Even in our telephone interview, Nicholas Wright, a close member of Shepard's London circle, had as many questions thirty years later about Shepard as I did.

The ambivalent nature of Shepard's persona extended to his work. As Shepard himself admitted "I never know what to say when somebody says what are the plays about. They're about the moment of writing."[14] This perhaps encouraged assumptions and interpretative projections that the Court chose to emphasise in their publicity, reflecting the Court's own Anglicised interpretation of American culture. The fact that the Court was fast becoming Shepard's theatrical home or that a Court press release for *The Unseen Hand* pitched Shepard as the most indicative voice of his generation suggests that what the Court was selling was not simply an up-and-coming American playwright, but more so a quintessential vision of America, despite the fact that Shepard's highly idiosyncratic writing is full of contradictions within itself, not to mention within the context of his counter-culture contemporaries.

I shall focus on four Royal Court productions of Shepard plays: *La Turista* (1969), *The Unseen Hand* (1973), *Geography of a Horse Dreamer* (1974) and *The Tooth of Crime* (1974). In addition to posters and visuals chosen to promote an image indicative of the spirit of the work, I shall refer to press releases that would have appeared in an entertainment column for the theatre-going public and predicated on the need for box office appeal. Such press releases were in turn filtered through newspaper columnists whose task it was to extract what was felt relevant (relevance in the press often being equated with the sensational). These press releases often use an isolated image as a hook to gain public attention, contributing towards the creation of a horizon of expectations that associates American culture with the supernatural, the grotesque, the primitive and the violent. Such publicity sold not simply Shepard's plays but American culture as a barbaric freak show, exaggerating the otherworldliness of this American vision to allay any British nationalistic chauvinism, fearing that such a cultural threat might implicate superior British audiences as equals.

I shall also consider the relationship between this promise to dramaturgically contain "barbaric" American culture within the theatrical event and subsequent critical response to the performance of Shepard's often enigmatic texts, which would have invited a host of meanings, including xenophobic ones. In two cases (*La Turista* and *Geography*) a lack of promotional sign-posts led to a critical backlash for productions that were seen as failing to realize their publicised promise of cultural containment, whilst there is a sense that the other two productions (*Unseen Hand*, *Tooth*) were greeted more favourably, possibly due to a greater appearance of such containment in the texts themselves and in performance. Nonetheless, not only was the barbaric nature of American culture an object of a critic's narrative interpretation, but Shepard's very dramaturgy attracted the label "primitive," something that may have influenced a response from Shepard in the writing of *Geography*, and its publicity.

Whilst considering critical reactions to performances, it must be emphasised that the reviews of a handful of theatre critics can constitute only a basic (and possibly uncharacteristic) snapshot of "audience" response, but in lieu of additional data such as post-performance questionnaires, they may offer at least a glimpse of the London response to Shepard's plays.

America the Primitive: *La Turista*

The first major Court production of a Shepard play was a revival of his earlier *La Turista*. It was the second major play to launch their newly opened Theatre Upstairs and directed by Roger Hendricks-Simon, one of a cohort of American expatriate theatre practitioners residing in 1970s London.

La Turista concerns a listless American couple, Kent and Salem, on holiday in Mexico, whose plastic lifestyles prove incompatible with the earthy culture surging around them (it is not without reason they are named after American cigarettes, indicative of the consumer culture that has burnt them out). The title *La Turista* is a pun both for "tourist" and a Mexican slang for dysentery that plagues the protagonists. The couple are invaded by foreign microbes causing the dysentery and simultaneously are presented as foreign bodies being purged from the host country. Later, at a dramaturgical level, Kent is purged out of the performance event altogether when he is transformed into a sort of Frankenstein monster, swinging on a rope above the audience and smashing his way through the set, but not before a well-intended Mexican Witch Doctor attempts to cure the couple by sacrificing a pair of hens in ritual slaughter, covering Kent and Salem in the birds' blood. Even though the Witch Doctor is presented as technologically primitive, the play questions if it is really Kent and Salem and their superficial values who are lower on the evolutionary ladder.

Whilst these characters might all be associated with both the supernatural and the primitive, the play itself has been often associated with a certain barbarism. The original New York production became infamous for its decision to use live hens every night for the duration of the play's run, and to slaughter them in front of the audience in real-time. All told, more than 40 hens lost their lives to *La Turista*. This ritual slaughter may have fed out of a general trend in the Off-Off-Broadway in which boundaries between theatre and ritual were being reassessed, but such actions were nevertheless viewed by animal rights activists as inhumane.

So too was the opinion in the London theatre scene. Hendricks-Simon was quoted in Royal Court publicity materials as saying "I don't think the British audience would stand for all those killings, and besides it would be very messy."[15] A directorial decision was taken to use live chickens in the production, however, at the last minute, through slight-of-hand, dummy hens would emerge whilst the genuine birds were spirited away to safety. Such previews read boldly "Reprieve for Witchcraft Hens" with photographs showing the pair of hens alive and well.

That this substitution of dummy for real hens was the focal point of several newspapers' angles suggests what values the Court assumed the public would attribute to American culture, evident in the other major newsbyte from the publicity material: "…a poetic and theatrical image of an undeniable here-and-now, uniquely American madness." [16]Another major photograph used was of the character of the Witch Doctor played by George Margo, a painted Native American Indian in the middle of his sacrificial dance. Whilst the play itself is suggesting that American culture might benefit from the wisdom of older "primitive" ones, the publicity simply conflates American and Mexican elements as one. It also stresses that the play "has an all-American cast", as though to forewarn (or promise) British audiences a glimpse at the primitive antics of American (and decidedly not British) culture.

The sales pitch pointed out that the measures taken to spare any animals of such a brutal demise risked potentially thwarting authorial intentions (in the face of the Court's reputation as a "writer's theatre"). The fact Hendricks-Simon felt the British audiences would not stand it implies perhaps that the British culture was too faint in the face of the savage American, even if it was a play that had been decided to be produced on a major London stage. It thus made a public statement about the limits of barbarism in Britain: ritual slaughter may happen "only in America" and thus the publicity encouraged a horizon of expectation not only selling the play's exoticism but the Court's ability to contain it in performance.

Perhaps because there were few dramaturgical signposts from prior Shepard productions, (as this was the first major one in London), there may well have been a sense that the exoticism was beyond containment, quite literally, in the play's final moments as Kent sailed over the heads of the audience and burst through the back wall of the set. The hens may have not lost their heads, but the audience may very well have. There is likewise in much of Shepard's work a rebellious stance towards the more Aristotelian unities of character and action, with characters often making abrupt emotional shifts amidst an illogical chain of events. Compared with much of what critics would have seen at the Court, such dramaturgical experimentation may have been mistaken as shoddy craftsmanship on Shepard's part.

Most reviews expressed a disdain for the play's slippery resistance to being "understood" and found the quintessentially Shepardian monologues tiring, perhaps most vocally when *Times* critic Irving Wardle disparaged the play on 19 March 1969 as "weak political satire" that paled in contrast with intellectual "political" Court writers. The fact that Shepard's work is seen as anti-intellectual seems to endorse Putzel's assertion of the British view of America as a savagely exotic phenomenon. The fact that this play was chosen out of what had already been a prolific body of his work up to this point (many of which do not feature ritual slaughter) and that its ritual slaughter found its way into the publicity limelight says something of the Court's vision of America the primitive. One must at least consider, at this early point in Shepard's relationship to his critics, the role of publicity in this cyclical reception of the plays, but more the horizon of expectations created by the wider context of writers more traditionally associated with the Court.

America the Alien: *The Unseen Hand*

The next Court production of a Shepard play was a revival of *The Unseen Hand* (1973), a glance at small-town America through the separate lenses of science fiction and the mythic American West. As with *La Turista*, there was a decision to publicise an image of America as both outsider and primitive. Its central character is an ape-like alien, Willy the Space Freak, who comes from the far reaches of the universe to a town named Azusa. Willy resurrects a gang of outlaw cowboys to overthrow a tyrannical interplanetary regime that have a psychic control over Willy, "the unseen hand." Later, a maniacally all-American male cheerleader appears, brutally holding Willy and the cowboys at gunpoint whilst he indulges in a patriotic homily to all things American. Ultimately, the play questions whether the inhabitants of Azusa are less civilised than this simian extraterrestrial, or any less the victims of social hegemony.

As the publicity emphasis for *La Turista* associated ritual slaughter with the barbaric and the supernatural, publicity for *The Unseen Hand* embodied a vision of America as primitive and yet foreign to the point of being unearthly. It further conflated the play's theme of cosmic invasion with an overview of the foreign talents infiltrating the Court, giving prominence to the non-British nationality of Australian director Jim Sharman and designer Brian Thompson, New Zealander Richard O'Brien starring as Willy, and American playwright Shepard.[17] Such publicity suggested the Theatre Upstairs was somehow being invaded if not by a force from the far reaches of space then by an alien closer to home.

In a personal interview on 9 October 2003, another Court designer, David Short, revealed that this emphasis on foreign intruders at the Court translated into a major publicity angle focusing on the centrepiece of the play's set, a 1951 Chevy in the diminutive Theatre Upstairs without clear access to load the piece. As Frank Lipsius said "The real mystery is how the car got up the steps to the theatre – or where it was found in the first place."[18] In an e-mail to the author on 25 February, 2005, Richard O'Brien recounted this publicity when he said:

> The Theatre Upstairs was remarkable for its ability to change its shape. There was a TARDIS-like quality to this very small rectangle of space that was emphasized leading to the run.

In addition to the car there was also the additionally incongruent appearance of a genuine lawn upon which it sat, "carried square foot by square foot up to the theatre."[19] Not only was there a space invader in the form of Willy the Space Freak, but an invasion of British theatrical space by ex-colonials, literally bringing their own turf into the Court.

This unearthly ship-in-a-bottle manifestation was not without connotations of violence and destruction. The Chevy was in fact a mangled shell, which director Jim Sharman remarked had been inspired by Andy Warhol's famous pop-art "Car Crash" as a visual influence for the production.[20] The threat of violent invasion was reinforced by the promotional poster's disturbing image, featuring an androgynous humanoid head, as though the head of an android, with its top neatly sliced off (Fig. 1).

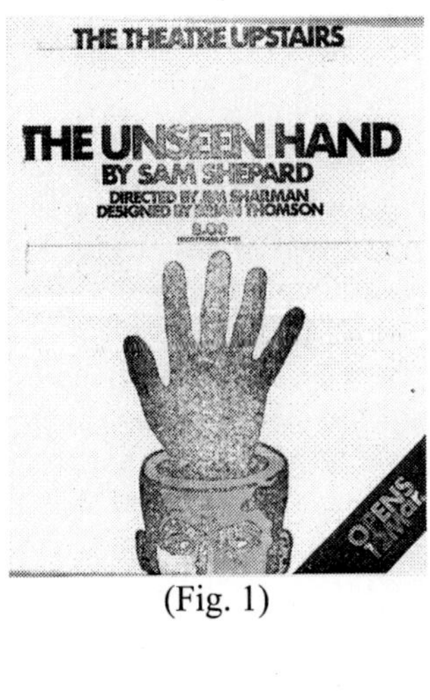

(Fig. 1)

The head in fact appeared as hollow as the wrecked hull of the Chevy, otherwise empty of human tissue, except for an enormous hand reaching up and out of the head, again pointing to a brutish corporeal presence secreted in a civilised (cerebral) space waiting to burst forth. One could argue that the threat of imminent alien invasion associated via publicity for *The Unseen Hand* contributed toward a horizon of expectation informed by xenophobia even more so than did *La Turista*. Yet in performance *The Unseen Hand* confounded xenophobic expectations, and in doing so may have contributed to a more positive critical reaction. A major device to assuage such tensions was the abrupt shift of action into musical numbers, with songs beckoning to a sanitized 1950s American culture, such as "Duke of Earl." There was also the presentation of characters as equally helpless victims incapable of launching an assault any more than, say, the characters found in the television series *Dad's Army*. Further, any potential menace of the play itself was undermined with more recognizably burlesque sight gags (as when the Kid tries to hold the cowboys at gunpoint with his trousers dropping). As O'Brien recounted:

> During the very successful run of *The Unseen Hand*, we had extra seats provided by the simple expedient of placing cushions on the floor. The "interaction" those nights – was laughter – joyous howlings of laughter.[21]

The critical reception reflected a comfort level not evident in *La Turista*. A typical review for *The Unseen Hand* was Robert Brustein's praise in the *Observer*, 18 March 1973: "Jim Sharman's production is splendid, a triumph for the fringe, and his actors have located precisely the right [inept] identities for their characters." Frank Lipsius hailed "this confrontation of incongruous funnies" with the characters' bewildered "blank looks (at which Warren Clarke is superb–no irony intended)"[22] There was also Michael Billingtons's observation in the *Guardian*, 13 March 1973 that Shepard "tells one more about the texture of American life than many of his contemporaries." That the text itself presents these characters collectively floundering more than those encountered in *La Turista* might have had something to do with this.

Nonetheless, criticism homed in on the combination of Shepard's vision as both primitive and otherworldly, with a tendency to suggest not only the culture represented within the world of the play as primitive and alien, but also the playwright himself. In the *Times*, 13 March, 1973, Irving Wardle emphasized the "unearthly" quality of the piece, yet saw Shepard as trying (and failing) to imitate Heathcote Williams' *AC/DC*, even though *The Unseen Hand* was written and first produced at La Mama, in 1969, and Williams' play almost a year later. Further, Wardle suggested that Shepard's play was just written for its London premiere at the Court, and that it

> ..has evidently received a massive shock from Heathcote Williams' *AC/DC.* At its worst, this has tempted [Shepard] into pseudo-scientific tirades about dimensional overlays and spatial projections; terms which Williams uses with understanding hit the *Dr. Who* level in Shepard. [23]

Again, there was a lack of information amongst the critics regarding Shepard's work in relation to other playwrights, and, as with *La Turista,* Shepard would be disparagingly contrasted with deft British writing, dismissing the primitive Shepard as no interloper into this territory.

A reviewer for the *Daily Telegraph* on 13 March 1973 indulged the tendency to associate the playwright with savagery and the extraterrestrial:

> We have it on good authority that the play *The Unseen Hand*...came out of the head of the American writer Sam Shepard...I am inclined to believe that, like the hybrid creature of baboon ancestry around whom or which the action bizarrely revolves, it [the play] came from outer space.

Billington likewise alluded to *Dr. Who* when placing both Shepard and the Space Freak as American, referring to O'Brien's "galactic visitor with a voice like a Brooklyn Dalek" and the play's lack of intellectual merit "Shepard may be no great thinker: but he is certainly a remarkable theatrical poet." Billington further partitioned the play's vision as reflecting only American culture when he added: "Now that Shepard is living in Britain it will be fascinating to see what he makes of our own bewildered, uncertain tinsel society." Billington would soon find out.[24]

America was in this case seen as the savage space-invader, a hulking wreck occupying turf whose vast size is matched only by the destruction it produces, yet one that easily contained by its own inferior pop-culture. Such a presentation of America within the world of the play was acceptable for the critics, who happily tarred Shepard himself with the same brush, equating him intellectually with the writing level of the low-budget children's show whilst being safely regaled with vintage pop-songs. This association reinforced and was in turn reinforced and possibly presaged by the Court's publicity of Shepard's work, having had its own unseen hand in equating Shepard's play with the trans-dimensional incongruities of the TARDIS and threatening pop-culture foreigners that are the stuff of only cheap science-fiction. This vision of America the alien, however, would change in Shepard's next play, which would re-focus any cultural critique previously reserved for America, and which would heighten any existing cross-cultural tensions and the already antagonistic relationship between playwright and critics.

America the Grotesque: *Geography of a Horse Dreamer*

Shepard would later reflect "*Geography of a Horse Dreamer* was written in London, and there's only one truly American character in the play."[25] It was indeed a response to expatriate anxieties. Its world premiere was the first Court production of a Shepard play written whilst he resided in London, one of the first commissions of their newly launched Neville Blond fund and also Shepard's directing debut.

Perhaps as both writer and director, Shepard assumed a greater degree of marketing control over publicity than in the two earlier Court productions. Given his own penchant for obscurity, both in his private life and attributing "meaning" to his plays, there was in the Court's publicity of *Geography* an attempt to conceal the play's narrative details, and by coincidence, this was the first (and last) major Shepard play to render British characters on a London stage.

The play concerns a young cowboy from Wyoming named Cody, whose ESP enables him to foretell the winning horses at local races. He is kidnapped by gangsters and held captive in a shoddy London hotel to do their bidding, but his surroundings only weaken his abilities until he is reduced mentally and physically to a dog. As with *La Turista* and *The Unseen Hand*, Cody is presented as both an invader (he is clearly a foreign body there) and invaded (the longer he stays in London the more mutilated his psychic powers become). His British captors are led by a foppish dandy named Fingers and a corpulent Doctor, the latter attempting to do a makeshift lobotomy on Cody, by ripping a magical "dreamer's bone" from the back of Cody's neck. The Doctor is stopped in the nick of time by the arrival of Cody's two brothers, who barge in without prior notice and mow almost everyone down with their shotguns to "rescue" Cody.

The text of *Geography* would itself both use then betray the horizon of expectations established in the Court's earlier publicity of Shepard productions. Obviously the play locates its otherworldly tensions not as originating from another planet or in Mexico, but in a country nonetheless foreign to its American protagonist (but not to its implied audience). The other major difference with the two plays prior is the presentation of foreigners. *La Turista* presents the Mexican Witch Doctor who slaughters hens as a ritual to cure Salem and Kent, and Willy the Space Freak is presented more a hapless victim than a monstrous space-invader. In *Geography*, however, the English Doctor is shown malevolently trying to destroy Cody, invading Cody's personal and psychic space with his incisions. The potential threat to Cody in the other two British characters, the effete Fingers and the fawning Waiter, is contained by the characters' ineffectual and emasculated natures.

Despite the presence of British characters, the Court's publicity of *Geography of a Horse Dreamer* played down any cross-cultural resonances. Court press releases stressed the play's subtitle "A Mystery in Two Acts", and offered the rather unassuming statement "It relates the adventures of a young American whose powers of ESP get him kidnapped by a gang of criminals."[26] By only suggesting that the story was about "a young American" and not revealing the nationality of the gang or the play's London setting, there was an implication the gang were likewise American by default. Even on 20 February, 1974, in an interview with Shepard the week prior to *Geography* opening, *Guardian* critic Michael White noted "Shepard is not keen to discuss the play" and quoted Shepard as saying "People always expect something when they come to the theatre. The more that expectation is left open the more chance there is that something can happen."[27]

Likewise, limited visual information trickled out. The main poster depicted a large nondescript hand pointing its index finger at one of several identical horses' heads, if anything, implying Walter Benjamin's notion of mechanical reproduction of images indicative of American disposable "MacDonald's" culture (Fig. 2). Another visual was a photograph of Buffalo Bill featured in the *Plays & Players* issue previewing *Geography*. Underneath the caption read "Images of the Wild West, a key element in the writing of Sam Shepard."[28] This, coupled with Shepard's growing London reputation as a purveyor of the weird and the wonderful, appeared to have promised yet another American freak show without any hint at cross-cultural anxieties that fuel the play's dramatic tensions.

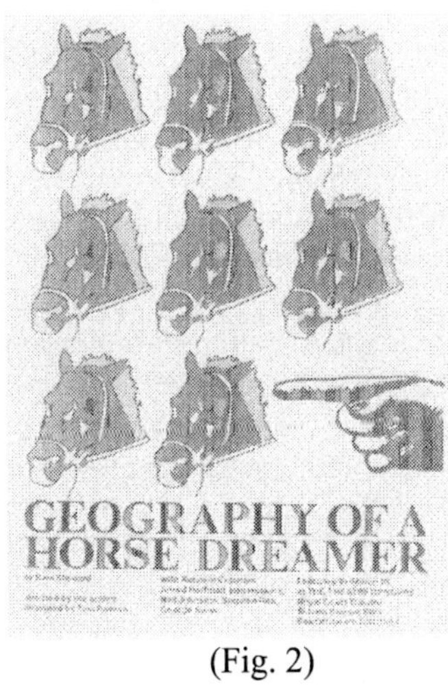

(Fig. 2)

The other major publicity image featured actor George Silver who played the sinister British Doctor. Images of Silver littered the press, referring not to his role but focusing on the man himself (Fig. 3). Court publicity clearly stated Silver had no prior acting training whatsoever, that this was his debut on the English stage and that Shepard cast Silver primarily on physical appearances. Silver's 27-stone figure became a sort of grotesque image of the larger-than-life quality of American culture itself, or at least, in the absence of any additional information, one might assume Silver represented this.[29]

(Fig. 3)

Only in performance would Silver's image be discovered to lampoon British culture, and only then in Act II, as the first act contained only Cody and American gangsters using a sort of American Runyonesque banter. In addition to the lack of reference to British culture in Court publicity, there was in the text an additional lulling the audience into believing that here again would be a freak show of barbaric American culture, as safely removed from England as Cody himself was removed from Wyoming. The trap sprang deftly on its audience in the second act, with the entrance of Fingers (presented in bowler hat, tweed trousers and matching cape, and with rings on every finger, including the thumbs), the Doctor, and the obsequious, cliché Waiter, who, apart from a token line such as "You rang, sir?" was as mute as a cigar-store Indian.

The unanimous critical response questioned whether the play "worked", with an interesting coincidence between reactions to the two acts. Act I was seen generally as "promising", but there was general disdain for Act II with the arrival of the British characters. *Punch* on 6 March, 1974 noted that "the creative spring all but goes haywire after the interval." John Barber of the *Daily Telegraph* seemed initially fond of the play. In a review on 22 February 1974 he wrote "It begins as a comedy thriller out of Damon Runyon. But it *loses its way*, and ends in Grand Guignol" (my italics). Up to the end of Act I, Barber quipped, "All this is about as enjoyable as a decent B-picture" before adding "But *it lost me* when a mad doctor turned up—a plan frustrated in a surrealist trick ending" (my italics). On 22 February, 1974 Times critic Irving Wardle decried Shepard's British characters as banal:

> Shepard's aim…to mix British elements into his dramatic crucible…In that area the play fails: not because the gentlemanly Fingers and the grotesquely corpulent Doctor are Hollywood projections of Britain, but because *their dialogue compares so lifelessly* with that of the American characters (my italics).

Even *Guardian* critic Michael Billington, often the staunchest of Shepard's London critics, concluded on 22 February, 1974 "it lacks the feeling you get in Shepard's better work that you are actively *participating in the modern American nightmare*" as though British existence could never be the stuff American nightmares were made of (my italics).[30]

As Putzel has pointed out, "In *Geography of a Horse Dreamer*…the British see themselves implicated and parodied."[31] The critics may very well have found the joke was on them, with the angle of the publicity (in this case what press releases did not tell), playing to a horizon of expectations that the play in its performance then undermined. What the London critics seem to have disdained was not so much Huck Finn offering a window overlooking the pond to that barbaric ex-colony, but rather a mirror, a threat that had infiltrated their own mechanism of the theatre previews column.

America the Rocky Horror: *The Tooth of Crime*

After the mixed reception of *Geography of a Horse Dreamer*, Shepard would never be in as high a position of power at the Royal Court, and in its wake he himself humbly conceded qualms about directing his own plays.[32] For the next Court production in 1974, a revival of *The Tooth of Crime*, Shepard's role was decidedly marginal. In an e-mail message, Music Director for the play, Richard Hartley, recalled that Shepard "…only came to a few *Tooth of Crime* rehearsals…I suspect he was a little perturbed by the rather grim way the play was developing."[33]

This choice to stage a revival of *The Tooth of Crime* so soon after its world premiere at the Open Space only two years prior was certainly a sign of Shepard's continued popularity at the Court. In equating Shepard's exoticism with this popularity, the Court quickly returned to publicity tactics that *Geography* had circumvented. Thus the Court both recycled its sales pitch of

American culture as barbaric, yet at the same time took a new tack that would draw from the horizon of expectations informed by critical responses to *The Unseen Hand.*

The Tooth of Crime has invited readings as diverse as the venerable maxim "the king is dead, long live the king" to being a battle between modernist and postmodernist forces, to a dirge for the death of rock. Its protagonist, Hoss, an amalgamation of American cultural icons (rock star, gangster, cowboy, to name a few) foresees his throne being usurped by a younger, more lethal rival. In Act II a character named Crow emerges challenging Hoss to a war of words and realizes Hoss' prophecy. The language itself is a dense metaphoric jargon with whiplash jolts of shifting syntax; he who controls those shifts controls the world of the play. In such a world, Hoss' final gesture of a gun in his mouth attempts to eradicate not only his own life but to acknowledge the futility of original expression within the textured web of language. Pulling the trigger concedes that words have literally failed Hoss, and yet even this dubiously original, speechless act is encoded in a series of repetition, where a number of Hoss' heroes have preceded him in suicide.

Whilst there is an undeniable air of violence to the text, it is a verbal violence with words used as weapons more than anything else, where language in the play takes on quite literally the effect of John Searle's speech acts. Yet the emphasis in the Court's publicity was of a more traditional sort, suggesting not the intellectual dexterity with which the play itself confronts and remoulds speech, with characters hurling words at one another and also at the audience. This linguistic violence was instead placed in mortal terms in the promotional poster, its typeset hearkening to that of a boxing advertisement, reading "THE THREE-ROUND ENTERTAINMENT BOUT OF THE CENTURY" (Fig. 4). Visually the blazing yellow poster was dominated by a pair of cowboys facing off in a "High Noon" desert shoot-out, one with a gun, the other wielding a guitar over his head, using the iconicity of the cowboy and the rock star to establish the American nationality of this violence.

(Fig. 4)

But at another level, the Court publicity attempted selling *Tooth* as a potential West End musical (no doubt instigated through Oscar Lewenstein's years as an impresario), spear-headed by the other major tag-line "the team responsible for both his earlier play *The Unseen Hand* and Richard O'Brien's *The Rocky Horror Show* are together again for this production."[34] This not only encouraged a confusion of genre, it associated the play's ethos with the otherworldly sci-fi camp of Richard O'Brien's work. Hartley pointed out, "The Court pushed the *Rocky Horror* connection in order to cash in on the success. With hindsight this was a big mistake":

> *Rocky Horror* was a musical, *Tooth of Crime* was a play with songs, in both cases the songs advanced the story and expanded on the emotional state of the characters. However, I think Sam would be horrified at any comparison 'cos I'm not sure he was a *Rocky* fan.[35]

In a similar e-mail interview O'Brien adamantly agreed:

> That publicity was very wrong—it drew comparisons where there were none and I think confused the audiences—"*Tooth of Crime*" is a bleak vision of the downside of fame and the isolation that occurs when you surround yourself with yes-men.[36]

In the end, the Court's production of *Tooth* was caught between competing signifying systems: it promised a visceral three-round American gladiator battle yet an all-singing, all-dancing spectacle comparable to contemporary West End rivals such as *Applause, Applause* or *Mame*. This publicity emphasis may well have been due to the promise of potential containment of American vulgarity that musical numbers had provided in *The Unseen Hand* (which was coincidentally produced by the "*Rocky Horror* team" as well).

What was interesting about the critical response to *Tooth*, however, was the greater willingness to accept it, although not for its series of monotonously dreary dirges masquerading as musical numbers. Whilst it remains tempting to suggest that this may have been because there were no British characters represented in this play (as there were in *Geography*), perhaps the critical acceptance of *Tooth* was partly due to the play's inevitable reliance on verbal over physical violence, even if the undeniably American trash-culture that informed this jargon might have seemed far more threatening to a British audience than that of physical force. An indicative line from the play reads "Got the molar chomps. Eyes stitched. You can vision what's sittin'. Very razor to cop z's sussin' me to be on the far end of the spectrum."[37] Such jargon assured Court audiences of an exotic culture with only nominal British connections, but more, that the potential physical violence promised in publicity was almost always contained within the play's more cerebral violence.

Reviews reflected a certain satisfaction at the play's linguistic vulgarity being further contained by the text's references to classical narrative conventions. In her review on 9 June 1974, Victoria Radin of the *Observer* considered it "basically a very simple play. It's even a universal one. The plot might come from Shakespeare." The *Guardian*'s Billington referred to the plot on 6 June 1974 as:

> …based on one of the most time-honoured dramatic myths…Despite its pounding rock score and often impenetrable jargon, Sam Shepard's *Tooth of Crime* strikes me as a deeply old-fashioned work that should go straight to the heart of all theatrical conservatives.

Even the *Times*' often acerbic Irving Wardle, in his review on 6 June, acknowledged the play's accessibility:

> The piece is quite classical in shape. The king learns his territory has been invaded and summons his advisers, a place and time are picked for the battle and courtesies exchanged; the contest then takes place with full medieval formalities, and the defeated king commits a heroic suicide.[38]

In this case, the critics were less praising *The Tooth of Crime* than the classical literary and dramaturgical tradition Shepard appeared to bow to and that seemed to contain the physical violence the play was promised to embody. The primitive American, at least in this instance, was almost tamed by the Bard himself.

Conclusion

By the time the next Court production of a Shepard play opened, the playwright would have returned to America. The Court's world premiere of *Action* (1974) may have found itself in a quandary, as this piece remains an anomaly not only amongst Shepard's work from his London period but within his entire oeuvre, and has come closest to being categorized "anti-American" in form and content. The piece was cautiously received as an attempt on Shepard's part to engage with and respond to the Absurdist "tradition", and the work of Samuel Beckett in particular. Not surprisingly, perhaps, *Action* got even less publicity than *Geography* received, other than to refer to it as "A play by Sam Shepard", and to focus on its four actors' previous work at the Court. [39]

The Court would return to its more traditional angle to sell the next Shepard play, the world premiere of his more naturalistic *Curse of the Starving Class*, written in the wake of his return from London and the first of what became known as his "family plays." It deals with the foreclosure of an American farm, whose mother wants to sell it in order to take the family to Europe, for "fine food, fine art, fine culture" without realizing her husband has already sold it and has been duped out of the money by loan sharks. This need to leave America comes to have fatal consequences for their teenage daughter, who is killed when a car bomb explodes as she herself tries to emigrate to foreign soil.

As with *La Turista*, the emphasis of the Court's publicity was a live animal onstage, this time a lamb. Though the animal itself was never actually at risk of being harmed, within the world of the play the lamb is brutally skinned alive. The play, it may be said, suffered the same fate. The caution *Action* received for its Absurdist allusions turned to outright dismissal when critics unanimously and vociferously viewed Shepard's incorporation of naturalism as a hackneyed effort to imitate his betters at writing the "well-made play." Director Nancy Meckler recounted "Sam phoned me that evening [of its opening night], and said not to blame myself or the cast. He said he was pretty sure the critics would act that way."[40]

This often subtle dialogue between the Royal Court and the London critical reception of Shepard's work points to an image of America as a primitive, violent, alien and inferior Other, admissible only within the safe confines of the arts, and even then with vigilance. Shepard may have had little control over what might have seemed a hermetically sealed system of discourse, one that he occasionally even played into (and, in rare cases, against) for professional promotion, but that remained both curious about and antagonistic towards his writing.

The closing words from *Curse of the Starving Class* may well have reflected Shepard's attitude to this intersection between the theatre, its critics and his own place somewhere in between. They may also be a cynical, if apt, meditation on the Anglo-American relationship he experienced during his London years, one that is reflected in the overriding image for this conference, the American eagle facing the British lion. Such a symbiotic yet destructive relationship is alluded to when the family's only two remaining characters, mother and son, recount a fable that has woven its way into their lives, that of a confrontation between a giant eagle and a very large cat. The eagle picks up the cat by accident, carrying it off into the sky, with each animal holding onto, yet trying to destroy, the other. As mother and son recite together:

> And they fight. They fight like crazy in the middle of the sky. That cat's tearing his chest out, and the eagle's trying to drop him, but the cat won't let go because he knows if he fall's he'll die…And they come crashing down to the earth. Both of them come crashing down. Like one whole thing.[41]

---

[1] Steve Putzel, "An American Cowboy on the English Fringe: Sam Shepard's London Audience", *Modern Drama*, 36.1 (1993): 131-146.
[2] Ibid, 133.
[3] Sam Shepard with Kenneth Chubb, "Metaphors, Mad Dogs and Old Time Cowboys." In *American Dreams*, edited by Bonnie Marranca. (New York, 1981), 200.
[4] Don Shewey, *Sam Shepard*, 2nd ed. (New York, 1997), 86.
[5] Personal email, 23 February 2006. Shepard & Chubb, "Metaphors, Mad Dogs and Old Time Cowboys," 201.
[6] Michael VerMeulen, "Sam Shepard, Yes, Yes, Yes", *Esquire,* 93, February 1980, 80.
[7] Personal interview, 23 July 2003
[8] Catherine Hughes, "New York Report", *Plays and Players,* May 1970, 16
[9] Personal interview, 14 March 2006; Clive Barnes, Theatre review for *Operation Sidewinder*, *New York Times*, 2 April 1970.
[10] Charles Marowitz, "A Sophisticate Abroad", *Village Voice*, September 1972, 59
[11] Shewey, *Sam Shepard*, 80; Marowitz, "A Sophisticate Abroad," 59; Nicholas Wright, telephone interview with the author, 22 January 2004; Peter Gill, telephone interview with the author, 14 January 2004.
[12] Max Stafford-Clark, telephone interview with the author, April 2006.
[13] Richard Findlater, ed. *25 Years of the English Stage Company*, (Ambergate,1981), 141.
[14] Walter Donohue, "American Graffiti: Walter Donohue on the Pulsating World of Sam Shepard", *Plays and Players*, April 1974, 17.
[15] Royal Court press release, *La Turista*, March 1969 (Courtesy of the Theatre Museum Study Room, London).
[16] Ibid.

[17] Royal Court press release, *The Unseen Hand*, 27 February 1973 (Courtesy of the Theatre Museum Study Room, London).
[18] Frank Lipsius, "The Unseen Hand", *Plays and Players,* May 1973, 48.
[19] Peter Ansorge, "The Wizards of Oz: Jim Sharman and Brian Thompson in interview with Peter Ansorge", *Plays and Players*, June 1974, 17.
[20] Ibid,14.
[21] Personal e-mail, 25 February 2005
[22] Lipsius, "The Unseen Hand", 48
[23] Irving Wardle, theatre review for The Unseen Hand, *Times*, 13 March 1973.
[24] Michael Billington, theatre review for The Unseen Hand, *Guardian,* 13 March 1973.
[25] Chubb, "Metaphors, Mad Dogs and Old Time Cowboys", 213.
[26] Royal Court press release, 1974 (Courtesy of the Theatre Museum Study Room, London).
[27] Ibid.
[28] Preview for Geography of a Horse Dreamer, *Plays and Players*, March 1974, 11.
[29] Royal Court press release, *Geography of a Horse Dreamer*, 1974 (Courtesy of the Theatre Museum Study Room, London).
[30] Theatre Review for Geography of a Horse Dreamer, *Punch*, 6, March 1974; John Barber, "Comedy Thriller Ends in Grand Guignol", *Daily Telegraph*, 22 February 1974; Irving Wardle, theatre review for Geography of a Horse Dreamer, *Times*, 22 February 1974; Michael Billington, theatre review for Geography of a Horse Dreamer, *Guardian,* 22 February 1974.
[31] Putzel, "An American Cowboy on the English Fringe", 137.
[32] Chubb, "Metaphors, Mad Dogs and Old Time Cowboys", 208.
[33] Personal email 16 April 2005.
[34] Royal Court press release, *The Tooth of the Crime*, 1974 (Courtesy of the Theatre Museum Study Room, London).
[35] Personal email, 16 April, 2005.
[36] Personal email 25 February 2005
[37] Sam Shepard, *Seven Plays* (New York: Bantam Books, 1981)
[38] Victoria Radin,Theatre Review for The Tooth of Crime, *Observer*, 9 June 1974; Michael Billington, Theatre Review for The Tooth of Crime, *Guardian*, 6 June 1974; Irving Wardle, Theatre Review for The Tooth of Crime, *Times*, 6 June 1974.
[39] Royal Court press release, *Action*, 1974 (Courtesy of the Theatre Museum Study Room, London).
[40] Nancy Meckler, personal interview with the author, 23 July 2003.
[41] Shepard, *Seven Plays*, 200.

CHAPTER FOURTEEN

RADIOHEAD'S AMERICA

JOSEPH TATE

As Virgil leads Dante under the words carved over Hell's gate—*Lasciate ogni speranza, voi ch'entrate*—they encounter first, on this side of Acheron, a collection of souls denied entrance to Hell yet unwanted in Heaven; souls the King James Bible refers to as the lukewarm: "because thou art lukewarme, and neither cold nor hot, I wil spew thee out of my mouth."[1] As Dante describes them, these innumerable, tepid souls lived *sanza 'nfamia e sanza lodo*: without infamy and without praise, neither cold nor hot.[2] They are the souls who, in Robert Pinsky's translation, were "neither rebellious to God nor faithful to Him;" souls who "Chose neither side, but kept themselves apart."[3] In the afterlife they waver between, eternally waiting in Hell's anteroom; taking no sides in life, they stand on neither side in death.

Academics often take sides, defending ideas and ideals. Think Noam Chomsky, Manny Marable, Edward Said, Cornel West, Howard Zinn, and others. And academic writing regularly handles ideological concerns like race, gender, and class, but criticism rarely claims that, as Frank Lentricchia wrote, "the point is not only to interpret texts, but in so interpreting them, change our society."[4] Some critics think this is the point of our activity, but hardly all; criticism's potentials vary as widely critics themselves and the texts they interpret. But criticism, defined here as a form of close reading—more simply, paying attention—can be political. This essay, in part, sounds the depths of criticism's political potential by exploring the music and art of Radiohead, an English music band whose work in various media implicitly incites its audience to critique their political world—to take sides and not remain lukewarm. But in close reading Radiohead, I wander Dante-like into dark woods, or the contemporary political turmoil surrounding and, in some eyes, created by American ascendancy.

Virtually no one today would question that, as Edward Said put it, "we live in a period of American ascendancy."[5] More specifically, we live in a period dominated by what emerged during Franklin D. Roosevelt's presidency and has increased exponentially in size since: the international influence of, to use Dwight D. Eisenhower's phrase, the American military-industrial complex or, to use Seymour Melman's words, America's permanent war economy.[6] While no one questions the fact of the American military-industrial complex's ascendance, people around the world question America's opinion of its geopolitical role. As Said wrote, America sees itself not as "a classical imperial power, but a righter of wrongs around the world, in pursuit of tyranny, in defense of freedom no matter the place or cost."[7] Said's description, though penned in 1993, summarizes the guiding principles of the current American administration's foreign policy as described in George W. Bush's second inaugural address on January 20, 2005:

> The survival of liberty in our land increasingly depends on the success of liberty in other lands. The best hope for peace in our world is the expansion of freedom in all the world. ... So it is the policy of the United States to seek and support the growth of democratic movements and institutions in every nation and culture, with the ultimate goal of ending tyranny in our world.[8]

Bush alone, however, cannot be credited with creating or maintaining this policy of "benevolent hegemony," as Robert Kagan terms it. Rooted in Truman's trepidations over communism's expansion, American interventionism has outgrown its beginnings to become the current "global war on terror."[9]

Though America's foreign policy agenda has been protean and persistent in equal measure, the war on terror differs so markedly in doctrine from prior American military actions that some scholars have called Bush's foreign policy revolutionary. America has changed from a country allied closely with other nations and international organizations (such as the United Nations and the North Atlantic Treaty Organization) to a country that jettisons "the constraints imposed by friends, allies and international institutions."[10] Yet in releasing itself from these constraints, the Bush administration has presided over a decline in America's international image. On August 15, 2001, one month before the September 11, 2001 terrorist attacks, the Pew Research Center's Global Attitudes Project found that people in Germany, France, Great Britain and Italy disapproved of George W. Bush's handling of international policy. Over seventy percent of interviewees from these countries agreed that "Bush makes decisions based entirely on U.S. interests, and most think he understands less about Europe than other American presidents."[11] Since that time, America's reputation has fluctuated with a general downward turn. Though some polls shows a positive upturn in image, a Pew report surveying 16 countries found nevertheless that, "the United States remains broadly disliked in most countries surveyed, and the opinion of the American people is not as positive as it once was."[12]

"Dislike" is an ambiguous, even lukewarm phrase and likely purposefully so. Yet, grasping instances of this dislike helps complete the picture of America's changing reputation. The music, art and public activism of Radiohead articulates a version of this dislike, but not as dislike for America *per se* or Americans. Instead, their work acknowledges the situation's complexity; they mount a sophisticated, allusive critique—sometimes explicit, sometimes implied—of America's global influence that sees America's new military posture not as a masquerade for imperialist pretensions and/or individual avarice, but as the result of a vast Piranesian network of seen and unseen pressures—economic, political, historical, and cultural.[13] To better hold the thread through this maze, a question needs answering: What or who is/are Radiohead?

Despite their relative global popularity—four of the band's six studio albums have reached RIAA certified platinum sales levels and they have fan web sites dedicated to them in Chinese, French, Greek, Italian, Japanese, Russian, and Spanish—it would be wrong to assume every reader of this collection knows that Radiohead is a five member experimental music group.[14] The band members, Colin Greenwood, Jonny Greenwood, Ed O'Brien, Phil Selway and Thom Yorke, met in 1987 at Abingdon's Boys School in Oxford, England and first named their rock group On A Friday. Renaming themselves after the song "Radio Head" from Talking Heads' 1986 album *True Stories*, the band evolved into their current manifestation: a music group that taps into influences ranging from post-bebop Jazz and avant-garde classical to techno, jazz and blues. In June 2003, Radiohead released their sixth studio album titled *Hail to the Thief (or, The Gloaming)*. Sonically less experimental than previous albums such as 2000's *Kid A* or 2001's *Amnesiac*, this album merged electronica with the traditional rock triumvirate of guitar, bass and drums.[15]

Overall an unsurprising, but hardly uninspiring, album, its title, however, disturbed some listeners in the United States. "Hail to the Thief" was a protest slogan in circulation following the 2000 American presidential election of George W. Bush over Albert Gore, Jr. The phrase puns satirically on the ceremonial tune "Hail to the Chief," a song also referred to as the "Presidential March."[16] In this election, contested popular votes from Florida were awarded to Bush by a Supreme Court vote of five to four. This gave Bush the electoral votes needed to win despite Albert Gore, Jr.'s plurality of the popular vote (50,999,897 to Bush's 50,456,002).[17] Voters dissatisfied with the election's outcome painted the phrase on signs to protest Bush's inauguration on January 20, 2001 in Washington, DC.

Given the phrase's recent and very political history, Radiohead's album title led to accusations of political extremism, accusations the band denied. Before the album's release in April 2003 readers of *New Musical Express* wrote in with comments. In a *NME* 5 April 2003 article written by Stephen Dalton, "How Radiohead's LP Title is the Biggest Anti-War Statement Yet," one respondent identified as "Kieran Evans" wrote:

> I believe in the US and I have faith in my president. I think it's sad this very talented band (perhaps the world's best) have gone from innovative lyrical and musical concepts to political protest. I don't need to get this CD. I think Jack White said it best: "I write songs about girls and being sad. Who am I to take a political stance?[18]

Another respondent identified as Ryan McGovern of Pennsylvania wrote:

> I'm a huge Radiohead fan but I'll be blacking out the title of the album when I buy it and ripping off the cover. I couldn't care less about their political beliefs. Celebrities are being mocked in America for screaming anti-war slogans because 70 per cent of the American people back the war.

An anonymous reader wrote: "Remember Thom, if it wasn't for us Americans and our "corrupt" government, you'd be speaking German right now!" I should add that these responses number among the less vitriolic; while some comments applauded the band's stance, most comments preferred a profanity-laced *ad hominem* approach.

These readers were responding less to the title itself than the intent *NME* claimed for it as "the Biggest Anti-War Statement Yet." But *NME* cannot be faulted for proposing such an interpretation: Radiohead's penchant—or at least Thom Yorke's penchant—for political statements is well known. Thom Yorke regularly and unambiguously dedicates songs to George W. Bush, and these dedications are rarely laudatory. At a concert in Bilbao, Spain, Alex Ross noted Yorke dedicated the song "No Surprises" to Bush, a song from their 1997 album *OK Computer,* predating Bush's tenure as president, with the lines "bring down the government / they don't speak for us," gaining new resonance.[19] Also considering that this was the band's first album recorded after September 11, 2001, and considering that America had invaded Iraq on March 19, 2003, only a month before the album's release, listeners were primed for controversy.[20] Yet in naming the album, the band followed the traditional procedure of extracting a specific lyric from the album itself: the title is a line from the first song "2+2=5 (The Lukewarm)." What interested listeners, reviewers and interviewers about the album title, however, was not how it threaded one song's thematic content throughout the entire album or how it refracted the song's relation to George Orwell's *1984.* Instead, people wanted to know how Radiohead intended to, as "No Surprises" says, bring down the government.[21]

In interviews before the album's release, Thom Yorke, the band's lead singer, claimed the title was not a reference to Bush's victory over Al Gore in

2000; Radiohead were musicians not politicians, they claimed.[22] As Chuck Klosterman notes in *Spin*:

> Since April [of 2003], Radiohead have stressed that *Hail to the Thief* is not a political record and that the album's title is not a reference to George W. Bush's controversial victory over Al Gore in the 2000 presidential election (in fact, Yorke claims he heard the phrase during a radio program analyzing the election of 1888).[23]

Around the same time, Jonny Greenwood—the quiet, eclectic musician at the band's centre who is now the BBC's composer in residence—wrote on Capitol Records' web site:

> We'd never name a record after one political event like Bush's election. The record's bigger than that. Hopefully it will last longer than Bush unless he's getting a whole dynasty together, which is always possible. One of the things Thom's singing about is whether or not you choose to deal with what's happening. There are a lot of lines about escaping and avoiding issues, about keeping your head down and waiting. Everybody feels like that from time to time as much as they feel frustration about things they can't change. It's a confusing time right now but that doesn't mean that we're issuing any kind of manifesto. It's more like we're summing up what it's like to be around in 2003.[24]

More explicitly, Yorke told *Time* magazine that *Hail to the Thief* does not limitedly refer to problems in the United States. The album title is, he claimed, "trying to express, without getting angry about it, the absurdity of *everything*. Not just a single Administration."[25] This strain was repeated again and again: "… with *Hail to the Thief*, the whole thing about it being political is a bit far-fetched. I keep reading stuff now about how this album is all about politics and anti-America … just because of the title and one or two quotes I gave. People overreact, read things into stuff, look for an angle…"[26] Yorke, no doubt, would say that is what I am doing here: this essay is looking for an angle and/or reading things, as he says, into stuff. But I would respond that he himself has given readers the angle. For instance, at the time of this particular interview, Yorke may have only given one or two quotes about Bush and American politics, but since then the number has risen, and continues to rise.

In November 2003, Yorke spoke out that Bush and Blair "... are not controlling the terrorist threat, they are escalating it. Blair will not be allowed off the hook by his pathetic pleading for us to 'move on' neither shall Bush."[27] At a nuclear disarmament rally in September 2004 Yorke said to the crowd that "we need to make it clear that we will not let America govern the world we live in."[28] America to the European is, Jean Baudrillard writes, a "form of interiorization of his or her own culture."[29] Given Great Britain's long-standing special relationship with America, as Howard Temperley explains, what the

English "have tended to see when they looked across the Atlantic have mostly been simplified models illustrating particular vices or virtues—the errors of kingship, the perils of democracy, the dangers of creeping socialism, or the iniquities of unrestrained capitalism."[30] This may be the case, but Yorke's vision of America seems less an imaginary interiorization—a misdirected projection of fears concerning England and/or the United Kingdom—than a real fear of a restless giant staring from the horizon's edge.[31]

What the lead singer says about Bush and America and what the album says about political influence, when compared, do not match up neatly. The band's music and art does not address the Bush Administration or America directly, yet the album does voice powerfully and perceptively the anxieties many people around the world have about American ascendancy or, if not American, the ascendancy of any political power that people follow without question. Still, at this point we might question: If *Hail to the Thief* is not a jab at Bush, then what if any purpose did the title have? Yorke has said:

> If the motivation for naming our album had been based solely on the U.S. election, I'd find that to be pretty shallow. To me, it's about forces that aren't necessarily human, forces that are creating this climate of fear. While making this record, I became obsessed with how certain people are able to inflict incredible pain on others while believing they're doing the right thing. They're taking people's souls from them before they're even dead. My girlfriend, she's a Dante expert, told me that was Dante's theory about authority. I was just overcome with all this fear and darkness. And that fear is the "thief."
>
> Fear and darkness. Living, but never *alive*; not living, just killing time—much like the lukewarm: "Hapless ones never alive."[32]

Yorke, in other comments seems to suggest he would rather sit in Hell's anteroom than dwell on politics:

> This record, to me, these new songs, they're not so much songs about politics as me desperately struggling to keep politics out. The past year … If I could have written about anything else, I would have. I tried really fucking hard. But how can any sensible person ignore what's been going on altogether? I couldn't, I really couldn't. Fuck, man, I would love to write lyrics free of politics![33]

Here Yorke confirms the album's grounding in the political present.[34] But why did (and do) Yorke and band members disavow politics as the album's basis? The question is not answerable with any definitiveness, but Yorke has provided hints. Whereas Ralph Ellison claimed he recognized "no dichotomy between art and protest," Yorke has said "I don't think there is much that's genuinely political art that is good art."[35] Yet, genuinely political and genuinely good art does exist, Yorke admits, citing Pablo Picasso's *Guernica* as an example.[36] I would add *1984*. I would add *Hail to the Thief*.

Hail to the Thief begins with "2+2=5 (The Lukewarm.)," and I reprint in full the song's lyrics from the album's CD sleeve:

Are you such a dreamer?
To put the world to rights?
I"ll stay home forever
Where two & two always makes up five

I'll lay down the tracks
Sandbag & hide
January has April's showers
And two & two always makes up five

IT'S THE DEVIL'S WAY NOW
THERE IS NO WAY OUT
YOU CAN SCREAM & YOU CAN SHOUT
IT IS TOO LATE NOW
BECAUSE
YOU HAVE NOT BEEN PAYING ATTENTION
I try to sing along
I get it all wrong
Eezeepeezeeeezeepeeezee
NOT
I swat em like flies but like flies the buggers keep coming back
NOT
Maybe not
"All hail to the thief"
"But I'm not"
"Don't question my authority or put me in the dock"
Cozimnot!
Go & tell the king that the sky is falling in
When it's not
Maybe not.

(ahh diddums.)[37]

The song's title derives from Orwell's *1984* and its content obliquely duplicates the book's themes.[38] The absurd mathematical formula, two plus two equals five, recurs throughout Orwell's book and captures what might be a, if not *the*, central lesson of *1984*: political pressures can shape our innermost selves.[39] Winston Smith, *1984*'s protagonist, is told by Julia, his lover, that the Party can never reach their essential selves: "They can make you say anything—anything—but they can't make you believe it. They can't get inside you."[40] Winston recalls these words as he traces the formula 2+2=5 on a table top at the book's end, sitting in the Chestnut Tree: “"They can't get inside you," she had said. But they could get inside you.”[41] As the anonymous Vogon guard bellows at Arthur Dent and Ford Prefect in Douglas Adams' *The Hitchhiker's Guide to the Galaxy*, “Resistance is useless.”[42] Thomas Pynchon writes in his forward to *1984*:

> The poor kid. You want to grab her and shake her. Because that is just what they do—they get inside, they put the whole question of the soul, of what we believe to be an inviolable inner core of the self, into harsh and terminal doubt. (xxiii)[43]

Orwell's book ends, however, on a positive note: written in the past tense, the appendix on Newspeak clarifies that the book's time period is over, even a resource for historical study.

Though evoking Orwell's *1984*, Radiohead's song speaks not from a particular character's voice, but in the voice of what we might call the lukewarm: an anonymous member of the book's society—the Party—someone content with ideology, that is, ideology as false consciousness. This song's speaker is not an active or sympathetic voice, but one removed from life and politics, someone happy to sandbag and hide, someone who follows without question. Tell them two plus two equals five and they believe it. But the song revels in contradiction and inconsistency. The narrator changes, signaled by punctuation marks above, but in the song Yorke modulates his vocal pitch and pacing and the song's multisectional structure, moving from 7/4 time to 4/4 time. The speaking voice from a deadened and relatively unsympathetic Winston-like character, one happy to stay home “where two & two always / makes up five,” to one more akin to O'Brien, *1984*'s antagonist, who admonishes repeatedly that we have not been “paying attention.” The inconsistencies mount: the phrase itself, “hail to the thief,” becomes the ideal example of doublethink. Unrestrained praise for the knowingly corrupt.

In reading the song this way, I abstract it. Chris Ott in his *Pitchfork* review writes that in this song, Yorke deals with his recent political distractions, pointing out the medieval ignorance of inaction in the face of overwhelming odds: “Are you such a dreamer/ To put the world to rights?/ I'll stay home forever/ Where two and two always makes up five.” It's a bit grandiose, but he rightly concedes the possible arrogance of his bravado during the tune's neurotically charged finale, “Go and tell the king that the sky is falling in/ When it's not/ Maybe not.”[44] Overall insightful, Ott's review locates the song in Yorke's universe, but given the song's refusal to stay put—its allusions to Orwell and Dante—linking the song to Yorke alone tends to cloister its power. The song's second word, “you,” could be self-directed, but if addressed to listeners, the implications magnify. We have not been, as the song tells us, paying attention.

To "pay attention" can mean any number of things, but given the song's title and sub-title, Orwell and Dante together would give us an image of active participation, the opposite of which means an afterlife in Hell's anteroom or eternal Party membership. A synonymic phrase from Orwell's writing for paying attention would be seeing what is in front of one's nose. Orwell wrote in 1946 that "To see what is in front of one's nose needs a constant struggle."[45] The four examples he gives elaborate a theory of what he calls the "power of holding simultaneously two beliefs which cancel out," a power that defines "our political thinking."[46] He also calls it the "all-prevailing schizophrenia of democratic societies, the lies that have to be told for vote-catching purposes, the silence about major issues, the distortions of the press."[47] His plural second-person pronoun casts a net wide enough to include Orwell himself and us. He elaborates how seeing what is in front of one's nose requires seeing past political deception, applying a form of common sense from one area of life to another: "When one is making out one's weekly budget, two and two invariably make four. Politics, on the other hands, is a sort of sub-atomic or non-Euclidean world where it is quite easy for the part to be greater than the whole or for two objects to be in the same place simultaneously."[48] Politics sustains a form of thinking where one can do the opposite of what one believes, where one can believe the opposite of what one does. Staying home, believing that two plus two equals five, despite knowing the formula is wrong: this is ideology.

Sonically, "2+2=5" could be called a more or less straightforward hard rock and roll song. The song begins with an intimate glimpse of the band: we first hear Jonny Greenwood plugging in a guitar during a studio sound check, the sounds we hear rumoured to be the first recorded during the album's sessions—a voice comments "That's a nice way to start Jonny." This moment of interrupted, fuzzed electric guitar gives way to a robotic (more likely computerised) drum-machine metronome and to a Radiohead signature sound: a clear-noted but melancholic Les Paul-esque jazz-guitar arpeggio.[49] About 1 minute 52 seconds into the song Phil Selway's real drumming replaces the computerised beats and Yorke's voice breaks loose, loud and firm to warn that we have not been paying attention. At almost exactly 2 minutes and 25 seconds (a possible numerical link to the song's title), the song changes again. Multi-sectional pop music is hardly new. Radiohead themselves attributed an earlier multisectional song, "Paranoid Android," as mimicking the Beatles "A Day in the Life." The song's greatest irony may be the fact it is straightforward rock-and-roll. The opening, staged or no, was no doubt used deliberately: an artifice to proclaim this is "real" rock, perhaps even in parody of rumours that, as Ott's review noted, this album would be a return to "rocking out." But in making this song a rock song, a song about refusal to question, much less refusal to rebel, the association of rock music with rebellion and resistance is overturned. This is a hard rock song about how hard it is to rock the boat.

Near the end of "2+2=5" the lyrics read: "I swat 'em like flies but like flies the buggers keep coming back." This line echoes Condoleezza Rice's quotation of President Bush in her prepared statement delivered to the 9/11 Commission on April 8, 2004:

> We also moved to develop a new and comprehensive strategy to eliminate the al-Qaida terrorist network. President Bush understood the threat, and he understood its importance. He made clear to us that he did not want to respond to al-Qaida one attack at a time. He told me he was "tired of swatting flies."

After her prepared statement, Rice was interviewed by the commission. Bob Kerrey asked as follows (from the transcript):

> MR. KERREY: Did – you've used the phrase a number of times, and I'm hoping with my question to disabuse you of using it in the future. You said the President was tired of swatting flies. Can you tell me one example where the President swatted a fly when it came to al Qaeda prior to 9/11?
> MS. RICE: I think what the President was speaking to was —
> MR. KERREY: No, no, what fly had he swatted?
> MS. RICE: Well, the disruptions abroad was what he was really focusing on.
> MR. KERREY: No, no —
> MS. RICE: When the CIA would go after Abu Sayyaf, go after this guy, and — that was what was meant.
> MR. KERREY: Dr. Rice, we didn't — we only swatted a fly once, on the 20th of August, 1998. We didn't swat any flies afterwards. How the hell could he be tired?
> MS. RICE: We swatted at — I think he felt that what the agency was doing was going after individual terrorists here and there, and that's what he meant by swatting flies. It was simply a figure of speech.
> MR. KERREY: Well, I think it's an unfortunate figure of speech...[50]

Though political, as the above shows, this song and the album overall sidesteps overt engagement: there are no references to specific events or people. However, the band's web site, which has long been an extension of the band's music, is not so shy. Following this album, the band's web site concentrated on corruption in economy and politics, but in the last year that focus has shifted more and more towards the most recent Iraq War and America's role in it. While still somewhat oblique, the web site displays an increasingly political, increasingly harsh critique of American power. The months before the American election in 2004 marked a decisive turning point in the imagery.

At first, the critique was somewhat cautious and subdued. For example, on November 3, 2004, the day after Bush's re-election the band's main web site displayed the image labeled Figure 1. The site's main page, which continues to change even as I complete this essay, showed this pill label (and I haven't tracked down the original medicine) along with its superadded caption. Oblique, perhaps. Subdued, perhaps. Yet this image had no other target than the American election. This is also an instance of what James Cody Walker has called "difficult laughter;" I laughed aloud to see this image dominating Radiohead's web site on 3 November. I laughed, but not long.

After Bush's re-election, the tone of the images changed dramatically. One image, a dark collage of George W. Bush's face and what appears to be a photo of a wide Los Angeles street. While Bush's face dominates the image, the pictured landscape has more vibrancy and close inspection reveals text take directly from *Why Do People Hate America?* by Ziauddin Sardar and Merryl Wyn Davies.

On their web site in early December 2004 a digital image with the file name "nicepeople.jpg," appeared. A collage: lines of small, black-and-white crying minotaurs stood alongside an oil-well pump photograph, both topped with pictures of George W. Bush and his first-term cabinet members. This collage might be better named a pastiche, in Fredric Jameson's use of the term. The top row displays American political leaders. Accompanying all of this lower right, there are Minotaurs standing in neat columns and rows.

Figure 1. All images are used by permission of the artists.
See: http://www.radiohead.com/offroad.html
[Despite the quality of this image, the author feels its inclusion is important.]

Figure 2*. Image file titled "fuckingbush.jpg"

Figure 3*. Image file titled "nicepeople.jpg"
At top from left to right we have, under separate headings:
Oil Industry, George Bush, President, Dick Cheney, Vice President, Donald Evans, Commerce Secretary[51], Condolezza Rice, National Security Advisor, Genetically Modified Foods, John Ashcroft, Attorney General[52], Donald Rumsfeld, Secretary of Defense, Ann Veneman, Secretary of Agriculture[53], Heavy Manufacturing, Paul O'Neill, Secretary Treasurer[54], Gale Norton, Secretary of the Interior.[55]

*[Despite the quality of these images, the author feels their inclusion is important.]

The list of members of Bush's first term administration (some have been replaced in Bush's second term), highlights their links to different economic sectors commonly under the eye of social and environmental groups. But, we might stop and ask about the juxtaposition: American leaders, an oil pump, and minotaurs. The obvious connection between the American leaders and the iconic oil pump, now synonymous with the Iraq war: America+oil=war in Iraq. The association of some cabinet members with their headings may, to some, be obscure. Condoleezza Rice (Secretary of State at the time of this writing) worked on Chevron's board of directors for ten years. But before departing Chevron and becoming National Security Advisor, Chevron named a 129,000 ton oil tanker after her.

> Q. Can I ask about Condoleezza Rice? Before she became National Security Advisor, she was on the Board of Directors of Chevron Corporation. And Chevron, before she left, named an oil tanker after her. There's an oil tanker named the Condoleezza Rice. And I'm wondering if—it's 136-ton oil tanker that carries oil around the world. And given that Chevron has been accused of human rights abuses with the Nigerian Mobile Police against civilians in Nigeria, I'm wondering whether the President thinks it's wise to have this close a relationship with Chevron.
> MR. MCCLELLAN. I think that issue has already been addressed by Dr. Rice, and she will uphold the highest ethical standards in office and that issue—
> Q. Should the President call the President of Chevron and say, take the name off the tanker?
> MR. MCCLELLAN. That issue has been addressed. I think the issue has been addressed.[56]

The American press devoured this, especially because at the time Chevron was involved in human rights abuses in Nigeria. The tanker's name was revealed in mid-April, but in early May 2001 it was renamed. The *SF Gate* reported:

> Leaving a wave of controversy in its wake, one of the most visible reminders of the Bush administration's ties to big oil—the 129,000-ton Chevron tanker Condoleezza Rice—has quietly been renamed, Chevron officials acknowledged yesterday.
>
> "We made the change to eliminate the unnecessary attention caused by the vessel's original name," said Chevron spokesman Fred Gorell. The double-hulled, Bahamian-registered oil tanker carrying the moniker of Bush's national security adviser was renamed the Altair Voyager, after a star, Gorell said.[57]

We might consider this throwaway trivia, but this constitutes one part of the larger process Radiohead's audiences often enter into: intensive reading of the band's music and art that leads into (sometimes requires) intensive reading of the world. Radiohead's fans have a unique tendency to decode the band's artwork, much as I am doing here. In essence, I am only doing what many fans already do. Gilles Deleuze: "The intensive way of reading, in contact with what's outside the book, as a flow meeting other flows ... getting [the book] to interact with other things, absolutely anything ... is reading with love."[58] The intensive reading that Radiohead's music and art attracts is part of Radiohead's fan culture. Many of Radiohead's fans—many, not all—dedicate themselves to investigating Radiohead's work as I have done. This essay, in short, does what fans do and often do better: read Radiohead.[59]

Zooming in (Figure 4) you see its a crying minotaur, often seen on the Radiohead web site and in album cover imagery. More often, the minotaur's hands are by its sides, as in Figure 5.

Figure 4. Detail of Figure 3, image file titled "nicepeople.jpg"

Figure 5. Image file name "minotaur2.gif"
Source: http://www.netsoc.ucd.ie/~johns/thin_ice/albums.html

We know it is a minotaur and not a devil, which it resembles, because the band has called it a minotaur. But why a minotaur? Joseph Halevi and Yanis Varoufakis call the U.S. economy "a latter day Minotaur single-mindedly concerned with its nourishment."[60] Applying Halevi and Varoufakis reading is tempting, but Stanley Donwood, the band's artist, writes on his web site:

> The minotaur, born of Pasiphae's illicit coupling with a white bull, was imprisoned in a labyrinth beneath King Minos' palace, and fed every seven years with teenagers sent as tribute from Troy. The half-human half-bull creature knew nothing but pain and humiliation, a monster created by humanity's fear of otherness, imprisoned in the dark catacombs beneath his mother's husband's palace. I thought of the monster, weeping, but not knowing how to feel regret. I don't know. It's hard to put this stuff into words.[61]

My earlier hunch proves to be an exemplary instance of "reading things into stuff," or overinterpretation.

Radiohead's minotaur mimics Bertrand de Jouvenel's minotaur, his metaphor for state power in the 1949 book *On Power: Its Nature and the History of Its Growth*.[62] The band uses the minotaur image to symbolize American political and economic power and its potential for abuse. But still why de Jouvenal's minotaur of all images? In this case, for Radiohead, the US minotaur is part bull, part man. The bull standing for the bull market America constantly chases. Also, the minotaur represents greed. The minotaur's origin deserves quick re-telling, even though Donwood's version might suffice. The minotaur resulted from the union of King Minos's wife and a bull, and the wife's erotic desire for a bull resulted from Minos' greed. Minos of Crete was sent a white bull by Poseiden to be sacrificed, yet Minos was so struck by the bull's beauty he switched the bull for another and sacrificed a different bull. Poseiden repaid this material lust by cursing his wife with erotic lust for the bull. The lesson: greed engenders monsters.

Orwell, across his writing, locates the solution to this political problem—how to resist ideology—not in governmental changes but in behavioral changes in the governed. Consider the animals in *Animal Farm*: they have seen, as the saying goes, the writing on the wall. They watch that writing change, degraded and erased, and choose inaction, trusting instead in memory's inevitable fault lines: perhaps they misremember.

Alex Ross, the *The New Yorker*'s music critic, has commented that "it's not very rock and roll to intellectualize rock and roll."[63] I could not agree more. Yet, Radiohead's music and art is hardly "rock and roll" as it has come to be understood; it's rarely about as Jack White said, "girls and being sad."[64] Consider this description of the band:

> They are noted in the music business for being polite and unproblematic. A

French hotel reservation of theirs extravagantly requested, "Extra towels." One hotel doorman described them as "nice, sharp-witted, entertaining young men. Not trash-headed and stupored, as you might expect."[65]

Add to this *Hail to the Thief*'s dedication: "To Patrick & Tamir and to a future worth having." Patrick and Tamir are the sons of Phil Selway and Jonny Greenwood respectively, and Yorke's son Noah may well be the addressee of the album's " Sail to the Moon. (Brush the Cobwebs out of the Sky.)":

Maybe you'll be president
But know right from wrong
Or in the flood you'll build an Ark
And sail us to the moon[66]

That the speaker wishes the addressee become president and not prime minister says much about the world that would be escaped by sailing to the moon.

In no uncertain terms, I read Radiohead's work as a politically charged call to action: the call comes from the description of the lukewarm, what one can become if not politically aware. Yet in attributing this call to Radiohead, I push the band into a role they would rather not play. Yorke feels uncomfortable, discouraged even, that people would look to his band for political motivation:

> the political framework itself is so dead and closed that people look to other sources, like artists, because art and music allow people a certain freedom. Obviously, the duty of artists is there, but it's more an indictment of the political system that someone like Zinn views artists as the seers, idealizing them as the people responsible [for inspiring] change. I think that would be great, but the reason people think like that is because there is no other element of participation anywhere.[67]

Yorke does not want to lead a revolution, but he has led protest rallies. Taken together, Radiohead's art and Radiohead's activism contradict: where there is a song decrying soul-crushing economic exigencies—"Dollars & Cents" on *Amnesiac* (notably not titled pounds and pence)—there is a musician objecting to contractual terms relating to internet downloads—Yorke posted (and later deleted) the following comments regarding the band's contract on Radiohead.com: "we would at least expect to get a few poultry pennies / if downloaded rather than nothing at all." Where there is a song wallowing in and warning of desire for false consciousness, there is an artist speaking out against global trade imbalance.

Raymond Williams surveyed Orwell's work and found when there is a "revolutionary socialist" there is also "the figure of the isolated, honest man who has seen beyond the socialist talk." Indeed, Williams writes, "the contradictions, the paradox of Orwell, must be seen as paramount. Instead of flattening out the contradictions by choosing this or that tendency as the "real" Orwell, or fragmenting them by separating this or that period or this or that genre, we ought to say that it is the paradoxes that are finally significant." And as Williams puts it, "some of the concepts we need for any full explanation may be beyond our reach just because of what we share with Orwell: a particular kind of historical pressure, a particular structure of responses and failures to respond."[68] Some of the concepts we need to explain Radiohead lie just our of reach precisely this same reason: we respond and fail to respond just as Radiohead themselves. We consider response and consider response the proper action, but our responses miss their mark or never arrive. Not all letters, not all post cards, arrive at their destination. There is, as Radiohead says, always a gap—a gap where dinosaurs roam the earth, where the sky turns green.

I want to end with a more obvious US-related image, this word-covered American flag, Figure 6.

Figure 6. Image file titled "punching.jpg"
[Despite the quality of this image, the author feels its inclusion is important.]

Unconnected in that they do not form a narrative, these words work against each other and together the words work against the American flag as icon. The words contort the flag's familiar design: stripes are reduced, stars are removed. The words use the flag as backdrop, revising the flag into a nonsensical sign and as a sign, the flag becomes an advertisement for words' power to contort and manipulate space, even the space of national identity.

Ultimately, Radiohead sees America not as an world ally but as a negative influence. Yet, their music and art suggests it is less capital-A America—the America Whitman found so impossible to define—than the Bush administration's version of America, one whose war economy and the globalization it advances are imagined as a devouring, raging minotaur. In Douglas Adams' *The Hitchhiker's Guide to the Galaxy*, a footnote describes the position of President of the Imperial Galactic Government: "The President in particular is very much a figurehead—he wields no real power whatsoever... For this reason the President is always a controversial choice, always an infuriating but fascinating character. His job is not to wield power but to draw attention away from it."[69] This quote encapsulates the problem of attributing American foreign or domestic policy George W. Bush alone. Bush, to be sure, wields power, but the network that supports his power is vast.

What Winston Churchill called the special relationship between England and America, Radiohead sees as dangerous and getting more so—in fact, Radiohead might claim the United Kingdom helps feed sacrifices to the minotaur. Radiohead's response to this? Music and art with knives out, work that imagines political circumstances in allusive terms that lead the listener past the lukewarm souls and deeper into the hell of problems, deeper to the centre where we will, like Dante, find our way out. The exit is unbeautiful; through Lucifer's matted hair and past icy-crusts, and past Winston, gin-scented tears trickling down the sides of his nose, believing 2+2=5. But once we are out, the heavens: full of stars.

---

[1] Dante Alighieri, trans. Robert Pinsky (London, 1994), III.9; *The Holy Bible* (London, 1611), sig. Z5[r].

[2] Dante, III.36.

[3] Ibid., III.33-34.

[4] Frank Lentricchia, *Criticism and Social Change* (Chicago, 1983), 10.

[5] Edward W. Said, *Culture and Imperialism* (New York, 1993), 285.

[6] Eisenhower coined the phrase in his "Farewell Radio and Television Address to the American People" on 17 January, 1961. See the *Public Papers of the Presidents of the United States: Dwight D. Eisenhower* (Washington, DC, 1961), 1035-1040.

[7] Said, 5.

[8] George W. Bush, 'President Sworn-In to Second Term' (inaugural address, 55th Inaugural Ceremony. Washington, D.C., 20 January 2005) http://www.whitehouse.gov/news/releases/2005/01/20050120-1.html.

[9] George W. Bush, 'President Addresses Military Families, Discusses War on Terror' (speech, Idaho Center, Nampa, Idaho, 24 August 2005. http://www.whitehouse.gov/news/releases/2005/08/20050824.html.

[10] Ivo H. Daalder and James M. Lindsay, 'Bush's Foreign Policy Revolution,' *The*

George W. Bush Presidency: An Early Assessment, Ed. Fred I. Greenstein (Baltimore, 2003), 100.

[11] Pew Global Attitudes Project, *Bush Unpopular in Europe, Seen As Unilateralist* (Washington D.C., 2001).
http://pewglobal.org/reports/print.php?ReportID=5.

[12] Pew Global Attitudes Project, *American Character Gets Mixed Reviews* (Washington D.C., 2005). http://pewglobal.org/reports/pdf/247.pdf (accessed September 15, 2005).

[13] See, for example, Ziauddin Sardar and Merryl Wyn Davies, *Why Do People Hate America?* (New York, 2003).

[14] Search for "Radiohead" at http://www.riaa.com/gp/database/ for record sales.

[15] Allan F. Moore and Anwar Ibrahim in '"Sounds Like Teen Spirit": Identifying Radiohead's Idiolect,' *The Music and Art of Radiohead*, ed. Joseph Tate (Aldershot, England: Ashgate, 2005) call this 'a problematic album, with its apparent refusal of musical and sonic progression, its partial recovery of guitars as sonic focus, and its re-employment of more conventional song structures' (139).

[16] The phrase "hail to the chief" derives from Sir Walter Scott's 1810 poem *The Lady of the Lake.* Adapted as a musical separately by two playwrights in the year it was published, Edmund John Eyre's version traveled to America carrying the tune now known as "Hail to the Chief."

[17] These numbers are taken from the Federal Election Commission's 2000 Official Presidential General Election Results page:
http://www.fec.gov/pubrec/2000presgeresults.htm.

[18] Steven Dalton. *NME* 2003, http://cobwebtheatre.com/html/hail_to_the_thief.html.

[19] Alex Ross. "The Searchers: Radiohead's Unquiet Revolution," *The New Yorker*. 116-17.

[20] *Amnesiac* was released after 11 September 2001 but that album was recorded during the *Kid A* sessions spanning late 1998, 1999 and early 2000.

[21] A question to ask at this stage: what other musicians have critiqued America in this way? The popular music's critique of American governmental may date at least as far back as Woody Guthrie's "This Land Is Your Land" with its original, lesser known verses:

One bright sunny morning in the shadow of the steeple
By the Relief office I saw my people—
As they stood there hungry, I stood there wondering if
This land was made for you and me.

See Mike Marqusee, *Chimes of Freedom: The Politics of Bob Dylan's Art* (New York, 2003). Despite the Reagan administration's adoption of Bruce Springsteen's "Born in the U.S.A.," the song critiques American domestic and foreign policy (this may be the supreme instance of what Adorno called regressive listening—failing to pay attention). The following list of critical songs is partial and reveals my limitations as a music listener: Blur's "Magic America," Steve Earle's "John Walker's Blues," Emenim's "White America," Green Day's *American Idiot,* John Lee Hooker's "This Land is Nobody's Land," Madonna's "American Life," Curtis Mayfield's "We the People Who Are Darker

Than Blue," Moby and Public Enemy's "Make Love Fuck War," Morrissey's "America Is Not The World," A Perfect Circle's "Counting Bodies Like Sheep to the Rhythm of the War Drums," Public Enemy's "Son of a Bush," Rammstein's "Amerika," System of a Down's "Boom!," Kanye West's "Jesus Walks," Wilco's "Ashes of American Flags." This list is far from comprehensive.

[22] Likewise, Orwell's readers wanted to read *1984* as a novel with specific targets. On 16 June 1949 he answered questions in a letter to Francis A. Henson of the United Automobile Workers:

My recent novel is NOT intended as an attack on Socialism or on the British Labour Party (of which I am a supporter) but as a show-up of the perversions to which a centralised economy is liable and which have already been partly realised in Communism and Fascism. I do not believe that the kind of society I describe necessarily *will* arrive, but I believe (allowing of course for the fact that the book is a satire) that something resembling it *could* arrive. ... The scene of the book is laid in Britain in order to emphasise that the English-speaking races are not innately better than anyone else and that totalitarianism, *if not fought against,* could triumph anywhere. (502)

[23] Chuck Klosterman, 'No More Knives' *Spin* 19.7 (July 2003): 66. What radio program this was, I have not been able determine. Further, I have found no evidence that the phrase 'hail to the thief' was used in the Cleveland-Harrison election contest, or the earlier contested 1876 Hayes-Tilden election.

[24] Asked if he wanted his second son Jeb (John Ellis) to run for president George H.W. Bush said, "Someday I would, yes." "Senior Bush plugs Jeb for president 'someday.'" CNN.com, http://www.cnn.com/2005/POLITICS/05/31/bush.plug/. Also see Kevin Phillips, *American Dynasty: Aristocracy, Fortune, and the Politics of Deceit in the House of Bush* (New York, 2004).

[25] Josh Tyrangiel. "How Radiohead Learned to Stop Worrying and Enjoy Being the Best Band in the World." *Time*. 161.23 (9 June 2003): 73.

[26] Neil Kulkami and Emma Morgan. "Alarms and Surprises." *Bang*. 4.3 (July 2003):60-71. 64).

[27] "Thom Yorke Leads Protest Over Bush Visit." *NME*. 19 November 2003.

[28] "Thom Yorke Leads 'Star Wars' Protest." *NME*. 27 September 2004.

[29] Jean Baudrillard, *America,* trans. Chris Turner (London, 1999), 75.

[30] Howard Temperley, *Britain and America since Independence,* (New York, 2002), 1. John Dumbrell writes in *A Special Relationship: Anglo-American Relations in the Cold War and After* (New York, 2001) that, "The term 'special relationship' appears to have been coined during World War Two" (7). Though in the 1960s to claim there was a special relationship "came to be regarded by Americans as something of a joke" (190). Temperley explains that "Economically, the two countries' affairs have been so closely intertwined that for much of the last two hundred years it has been more accurate to think of them as constituent parts of a single transatlantic enterprise than as two separate economies." (5).

[31] You don't really want a monster taking over. Tip toe around, tie him down. We don't want the loonies taking over.

[32] Dante, III.54.
[33] Kulkami and Morgan, 64.
[34] These words held up by Stanley Donwood during the band's 18 December 2002 web cast give a pointedly political cast to the album's title: "If you're a proper fucking thief you can't be fucked with burgling properties in the middle of the fucking night because you got to fucking get up in the fucking morning to run the fucking country."
[35] Ralph Ellison, "The Art of Fiction: An Interview," *The Collected Essays of Ralph Ellison*, ed. John Callahan (New York, 1995), 210-224: 211. Thom Yorke, Interview, "Duty of Expression," http://www.resonancemag.com/feature_01.html.
[36] He qualifies the example, however. The political impact of art, in Yorke's view, is only as strong as the art's impact *qua* art. Picasso's work succeeded politically because Picasso succeeded aesthetically. This, for Yorke, explains why people discuss the political ramifications of his band's work: if Radiohead's music was not successful *qua* music, its politics would not be under discussion.
[37] Radiohead, *Hail to the Thief, (or, The Gloaming.)* (Capitol, 2003). Here and in other instances I retain the album's punctuation. Aspects of this song not addressed here, such as the devil reference, are taken up in my essay "*Hail to the Thief:* A Rhizomatic Map in Fragments," *The Music and Art of Radiohead*, ed. Joseph Tate (Burlington, VT: Ashgate, 2005). That essay, like this one, however, can make no claim of comprehensiveness.
[38] The band's interest in Orwell's work extends also to the *Kid A* song "Optimistic" which refers to Orwell's *Animal Farm*. Also, their current web site's archive section is titled "The Memory Hole."
[39] Orwell uses the formula in an earlier essay, as I cite below. Though used to make an entirely different point, the formula also occurs in the first part of Fyodor Dostoyevsky's 1864 novella *Notes from the Underground.*
[40] George Orwell, 1984 (London, 1949), 170.
[41] Ibid., 300-301.
[42] Douglas Adams, *The Hitchhiker's Guide to the Galaxy* (New York, 1995), 69. The band named the second song on OK Computer, "Paranoid Android," after Marvin, the novel's morose robot.
[43] Thomas Pynchon also interests the band. Their self-run merchandise web site is titled W.A.S.T.E. after the underground postal system in Pynchon's *The Crying of Lot 49.*
[44] Chris Ott, Review of *Hail to the Thief, Pitchfork,* http://www.pitchforkmedia.com/record-reviews/r/radiohead/hail-to-the-thief.shtml.
[45] George Orwell, "In Front of Your Nose," *The Collected Essays, Journalism, and Letters of George Orwell*, ed. Sonia Orwell and Ian Angus (Boston, 2000) v.4, 125.
[46] Ibid., 123.
[47] Ibid., 124.
[48] Ibid., 125.
[49] This guitar sound owes a debt to the Smith's former guitarist Johnny Marr.
[50] 9-11 report.
[51] Donald Evans retains office as 34th secretary of U.S. Department of Commerce.
[52] As a Missouri Republican Senator, John Ashcroft advocated for genetically modified

foods. Ashcroft received $10,000 for his 1990s Senate bid from Monsanto, a large agricultural company focused on biotechnology. Rumsfeld was president of Searle Pharmaceuticals when it was acquired by Monsanto.

[53] Ann Veneman held a position on Calgene Pharmaceutical's board of directors, another company now owned by Monsanto, the first company to market with a genetically modified food in 1994: it was a gene-altered, better-ripening tomato, the rot-resistant Flavr Savr tomato. The US FDA announced on May 16, 1994 the tomato was safe for consumption. Yet, it flopped. As well as the safety concerns, they cost about twice as much as non-GM tomatoes, had no better flavour, and were prone to bruising. From "Flavr Savr Tomato & GM Tomato Puree: The Failure of the First GM Foods," *Soil Association*. 28 April 2005.
http://www.soilassociation.org/web/sa/saweb.nsf/Library?OpenForm&Cat=_Briefing_Sheets.

[54] Paul O'Neill resigned on Dec 6, 2002. Many news papers report the resignation as a "firing" over O'Neill's tax-cut reservations. A book by Ron Suskind on O'Neill is titled: *The Price of Loyalty: George W. Bush, the White House, and the Education of Paul O'Neill.* His biography on the United State Treasury web site reads: O'Neill was chairman and CEO of Alcoa from 1987 to 1999, and retired as chairman at the end of 2000. Prior to joining Alcoa, O'Neill was president of International Paper Company from 1985 to 1987, where he was vice president from 1977 to 1985.'
http://www.ustreas.gov/organization/bios/oneill-e.html.

[55] Norton was senior attorney for the Mountain States Legal Foundation, a group that defends mining, timber, and oil interests from environmental regulations. Its funding has come from wealthy conservatives and corporations such as Chevron, Texaco, Exxon, and U.S. Steel, among others. She was also a "lobbyist for NL Industries, the "L" stands for lead. NL Industries has been a defendant in lawsuits involving 75 toxic waste sites." http://www.pbs.org/now/transcript/transcript222_full.html.

[56] Scott McClellan, Press Briefing. *The White House* 2 April 2001. 28 April 2005. http://www.whitehouse.gov/news/briefings/20010402.html.

[57] Carla Marinucci, "Chevron Redubs Ship Named for Bush Aide Condoleezza Rice Drew Too Much Attention," *SFGate.com*. 5 May 2002, accessed 28 April 2005.
< http://sfgate.com/search/>.

[58] Gilles Deleuze, "Letter to a Harsh Critic" *Negotiations: 1972-1990* (New York, 1995), 8-9.

[59] A glance through the At Ease message board, http://ateaseweb.com/mb/, and Mortigi Tempo, the Green Plastic Radiohead message board, http://www.mortigitempo.com/, proves the point.

[60] Halevi, Joseph and Yanis Varoufakis. "The Global Minotaur" *Pox Americana: Exposing the American Empire.* Ed. John Bellamy Foster and Robert W. McChesney. (New York, 2004), 94.

[61] http://shop.slowlydownward.com/Store/DisplayIndividualItem/1/552.html

[62] Bertrand de Jouvenel, *On Power: Its Nature and the History of Its Growth.* Trans. J.F. Huntington. (New York, 1949).

[63] Alex Ross, "Rock 101," *The New Yorker,* 2003.
http://www.newyorker.com/critics/music/?030714crmu_music

[64] A song that comes closest to this description might be "Morning Bell" on *Kid A*, a song about divorce.

[65] Ross, "The Searchers," 111.

[66] Radiohead, *Hail to the Thief, (or, The Gloaming.)* (Capitol, 2003).

[67] Sarah Burton, Interview with Thom Yorke and Howard Zinn.
http://www.resonancemag.com/feature_01.html

[68] Raymond Williams, *George Orwell* (New York, 1971), 88-90.

[69] Adams, *The Hitchhiker's Guide*, 38.

FURTHER READING

Ephraim Douglass Adams, *Great Britain and the American Civil War* (London: Longmans, 1925).

D. Armitage, *The Ideological Origins of the British Empire* (Cambridge: Cambridge University Press, 2000).

Malcolm Bradbuy, *Dangerous Pilgrimages: Trans-Atlantic Mythologies and the Novel*, (London: Penguin Books, 1996).

Duncan Andrew Campbell, *English Public Opinion and the American Civil War* (Woodbridge, Suffolk: Boydell Press, 2003).

Martin Crawford, *The Anglo-American Crisis of the Mid-Nineteenth Century: The Times and America, 1850-1862* (Athens & London: The University of Georgia Press, 1987).

John Dumbrell *A Special Relationship: Anglo-American Relations in the Cold War and After* (New York: St. Martin's, 2001).

J.H Elliott, *The Old World and the New, 1492-1650,* (Cambridge: Cambridge University Press, 1992).

A.Fitzmaurice, *Humanism and America: An Intellectual History of English Colonisation 1500-1625* (Cambridge: Cambridge University Press, 2003).

Robert Bechtold Heilman, *America in English Fiction 1700-1800: The Influences of the American Revolution* (Baton Rouge: Louisiana State University Press, 1937).

Karen Ordahl Kupperman, ed., *America in European Consciousness* (Chapel Hill: University of North Carolina Press, 1995).

by Fred M. Leventhal and Roland Quinault, eds., *Anglo-American Attitudes: From Revolution to Partnership* (Aldershot: Ashgate, 2000).

Deborah Madsen, *American Exceptionalism* (Edinburgh: Edinburgh University Press, 1998)

Allan Nevins, *America through British Eyes: American Social History as Recorded by British Travellers* (Gloucester, MA: P. Smith, 1969),

T. Scanlon, *Colonial Writing and the New World 1583-1671: Allegories of Desire* (Cambridge: Cambridge University Press, 1999).

Daniel Snowman, *Kissing Cousins: An Interpretation of British and American Culture, 1945-1975* (London: Temple Smith, 1977)

Howard Temperley, *Britain and America since Independence,* (New York: Palgrave, 2002)

Francis Williams, *The American Invasion: On American Penetration of British Economy* (London: Anthony Blond, 1962).

List of Contributors

Catherine Armstrong is a teaching fellow at the University of Warwick where she completed her PhD in 2004. Her first monograph, *Writing North America in the Seventeenth Century* is due out shortly with Ashgate. Her research focuses on the representation of America in print during the early modern period.

John Barrington was born in New York City, but grew up in England, where he attended St. John's College, Oxford as an undergraduate. After a stint as a Chartered Accountant, he returned to History, receiving a Ph.D. from the College of William and Mary in Williamsburg, Virginia in 1997. He now teaches colonial American history at Furman University in Greenville, South Carolina.

David Brown is a Lecturer in American Studies at Manchester University. He is author of *Southern Outcast: Hinton Rowan Helper and The Impending Crisis of the South* (2006) and co-author (with Clive Webb) of *Race in the American South: From Slavery to Civil Rights* (2007).

Ian Chambers recently gained his doctorate in Native America/Early American History from the University of California, Riverside. He has been the recipient of fellowships from the John D. Rockefeller Library, Colonial Williamsburg and the Institute for Southern Studies at the University of South Carolina. His work focuses on the late seventeenth and early eighteenth century Atlantic world, with particular reference to the American southeast.

Roger Fagge teaches history at the University of Warwick. He has published on US and British social and cultural history, including the comparative study *Power, Culture and Conflict in the Coalfields: West Virginia and South Wales* (Manchester, 1996). He is currently completing a study of J. B. Priestley, alongside working on the cultural and political interaction between radical England and America in the twentieth-century.

Robert Grant completed his PhD in English Literature at the University of Kent in February 2004, where he also taught American Literature. During 2005, he was Visiting Fellow at the Institute of Commonwealth Studies in London and is currently an independent scholar. He published *Representations of British emigration, colonisation and settlement c.1800-1860* (2005) and regularly presents papers in the United Kingdom, the United States, Australia and New Zealand.

Tammy Grimshaw is the author of *Sexuality, Gender, and Power in Iris Murdoch's Fiction* (2005). She has written many articles on post-war British fiction and is currently teaching at the University of Sunderland.

Sarah Irving is a Junior Research Fellow in History at Wolfson College, Oxford University. She completed her Ph D at King's College, Cambridge. Her book, *Natural Science and the Origins of the British Empire* will be published by Pickering and Chatto, London, in 2008.

Tim Lockley is an Associate Professor of History at the University of Warwick. His publications include *Lines in the Sand: Race and Class in Lowcountry Georgia, 1750-1860* (Athens, Ga., 2001) and *Welfare and Charity in the Antebellum South* (Gainesville, Fl., 2007).

Allen McLaurin, formerly Head of the Department of Humanities, is currently Senior Academic in Humanities in the Faculty of Media and Humanities at the University of Lincoln. He has a particular interest in word and image studies, and his most recent research project is a study of British representations of America in the middle years of the twentieth century.

Brook Miller is an assistant professor of English at the University of Minnesota, Morris. His research focuses upon transatlantic studies and modern British literature.

Max D. Stites is currently in the final stages of completing a PhD in English at the University of Exeter, England. His essay "Hunter S. Thompson's Fiendish Vision of the American Dream" will be included in *American Dreams: Dialogues in U.S. Studies,* ed. Ricardo Miguez (Cambridge, 2007).

Timothy Rogers is completing a PhD in Drama at the University of Manchester, where he also contributes to undergraduate courses on postwar American and European theatre. His research focuses on playwright Sam Shepard's dramaturgical development during his expatriate "London" years between 1971-1974.

Stefan Schwarzkopf is a Lecturer in Marketing, Queen Mary College, University of London. He is currently finishing his PhD thesis at Birkbeck College on the politics and cultures of advertising in early twentieth-century Britain. His publications include an edited book on the political marketing of the German Social Democratic Party between the 1960s and the 1990s (Leipzig, 2003), and an article in *Contemporary British History* (June 2005) on the cross-links between the cultures of the Cold War and the advertising industry in the United Kingdom.

Joseph Tate is an independent scholar who lives and works in Missoula, Montana. He has published and lectured on Radiohead, Shakespeare and other topics.

David Waller is a Senior Lecturer in American Politics at the University of Northampton. He is currently researching Anglo-American diplomacy in the decade before the American Civil War and its relationship to contemporary racial and imperial ideologies